Principles of Nuclear Science and Engineering

RESEARCH STUDIES IN NUCLEAR TECHNOLOGY

Series Editor: **Jeffery D. Lewins DSc(Eng), CEng**
Engineering Laboratories, University of Cambridge
Fellow of Magdalene College, Cambridge, England

1. A Dictionary of Nuclear Power and Waste Management
   with Abbreviations and Acronyms
   Foo-Sun Lau
2. Radioactivity and Nuclear Waste Disposal
   Foo-Sun Lau
3. Principles of Nuclear Science and Engineering
   A. A. Harms

Principles of Nuclear Science and Engineering

A. A. Harms
McMaster University
Hamilton, Canada

RESEARCH STUDIES PRESS LTD.
Letchworth, Hertfordshire, England

JOHN WILEY & SONS INC.
New York · Chichester · Toronto · Brisbane · Singapore

RESEARCH STUDIES PRESS LTD.
58B Station Road, Letchworth, Herts. SG6 3BE, England

Marketing and Distribution:

Australia, New Zealand, South-east Asia:
Jacaranda-Wiley Ltd., Jacaranda Press
JOHN WILEY & SONS INC.
GPO Box 859, Brisbane, Queensland 4001, Australia

Canada:
JOHN WILEY & SONS CANADA LIMITED
22 Worcester Road, Rexdale, Ontario, Canada

Europe, Africa:
JOHN WILEY & SONS LIMITED
Baffins Lane, Chichester, West Sussex, England

North and South America and the rest of the world:
JOHN WILEY & SONS INC.
605 Third Avenue, New York, NY 10158, USA

Library of Congress Cataloging in Publication Data

Harms, A. A.
Principles of nuclear science and engineering.
(Research studies in nuclear technology ; 3)
Bibliography: p.
Includes index.
1. Nuclear engineering. I. Title. II. Series.
TK9145.H34 1987 621.48 87-9876
ISBN 0 86380 057 2
ISBN 0 471 91628 5 (Wiley)

British Library Cataloguing in Publication Data

Harms, A. A.
Principles of nuclear science and
engineering. —— (Research studies in
nuclear technology series).
1. Nuclear engineering
I. Title II. Series
621.48 TK9145

ISBN 0 86380 057 2

ISBN 0 86380 057 2 (Research Studies Press Ltd.)
ISBN 0 471 91628 5 (John Wiley & Sons Inc.)

Printed in Great Britain by SRP Ltd., Exeter

PREFACE

Nuclear science and engineering may be defined as the scientific-technological domain of theory and practice concerned with the management of matter-energy transformations. Its essential focus is, therefore, directed towards nuclear processes and their implications. The controlled release of energy from nuclear sources and its conversion into electricity has emerged as the dominant theme in this human endeavour.

Analogous to other academic-professional disciplines, nuclear science and engineering possesses its own history of intellectual thought and technological development. Unlike most other disciplines, however, it has emerged with remarkable rapidity and, by virtue of its intrinsic technological potential, has imposed itself with considerable impact. Institutions of higher learning and centers of research and development have thereby been subjected to several severe impositions. Among these is the need to educate specialized personnel and to develop a variety of devices in order to implement technological constructs into a complex industrial infrastructure under socially acceptable terms. Concurrently, there exists the compelling requirement to formulate a stimulating instructional academic framework to strengthen the learning foundations of this enterprise. The associated tensions demand a continuous recasting of nuclear science and engineering as a learning discipline.

In this text we seek to place the "introduction" of nuclear science and engineering into an instructional context by an emphasis on two important pedagogical considerations. One is that the subject may be effectively communicated by a "bottom-up" approach in which one seeks to impart knowledge and understanding by stressing continuity in relation to preceding course material; this contrasts to the "top-down" approach involving the simplification of a complex and specialized subject by a variety of conceptual reductions and analytical approximations. The other consideration emphasized here is based on the recognition that nuclear science and engineering is linked at a most basic physical level to time-varying processes of matter and energy transformations which can be efficiently introduced using dynamical descriptions.

Accordingly, we begin with a discussion and analysis of the fundamental connection between matter and energy. We then introduce selected nuclear phenomena stressing those features which are of particular relevance to nuclear engineering. The individual chapters dwell on specific themes with a focus towards technological relevance of nuclear science. With few exceptions, formulae are derived from first principles and a consistent notation is used throughout. As appropriate for an introduction to a subject for which there exists a large body of accumulated experience, we have endeavoured to stress the assumptions underlying the pedagogical constructs and to delineate their limits. Then, in order to enhance the student's learning experience, we have worked out some problems in an informal handwritten format. Selected data, physical constants, specialized topics, recommended demonstration experiments, and a reading list are found in the Appendices. Finally, a Chart of the Nuclides is located inside the back cover.

The prerequisite academic level assumed here is a solid understanding of the main areas of physical science and mathematics generally available at the beginning second year level in programs of science and engineering. The reader will soon recognize that this text is characterized by expository compactness, integrative breadth of material, and intuitive evolvement of a subject; this approach has been deliberately chosen because it appears particularly effective for the student population we have in mind.

This text has emerged over a number of years in the teaching of a course on the subject. Its format and content has benefited from the comments of numerous students as well a from a critical review of all or part of it by the author's associates, Professors John Cameron, Bill Garland, John Harvey and Dick Tomlinson, as well as his teaching assistant, Gerry Gaboury. As happens so often, direct and indirect stimulation for the evolvement of a particular approach can here also be traced to a number of personal contacts sustained over time; among those to be acknowledged in particular are Professors Freeman Dyson (Institute for Advanced Study, USA), David Goodings (McMaster University, Canada), Wolf Haefele (Kernforschungsanlage Juelich, FRG), George Miley (University of Illinois, USA), and Klaus Schoepf (University of Innsbruck, Austria). Specialized services were undertaken by the McMaster Engineering Word Processing Centre (typing), Christine Leng (drawings), Shirley Williams (text review), and Barry Diacon (class demonstration experiments). Finally, Professor Jeffery Lewins (University of Cambridge, UK) provided expert editorial counsel. To all, a sincere "Thank you!".

A.A. Harms
McMaster University, Hamilton, Canada
March 1987

To: Trudy, Lorchen, and Theodore

TABLE OF CONTENTS

TABLE OF CONTENTS (cont'd.)

TABLE OF CONTENTS (cont'd.)

CHAPTER I

MATTER AND ENERGY

Nuclear science and engineering is concerned with processes and applications of nuclear phenomena. Fundamental to an understanding of this subject is the relationship between matter and energy. Our intent in this first chapter is to highlight these connections at several levels of interest.

1.1 Our Physical World

Humans have always sought to understand their world. Processes in nature possessed a particular appeal and prompted an inquiry on topics such as the dynamics of the universe and the cyclical pattern of plant growth. Even such practical considerations as physiology and the structural strength of shelters were of interest. Nature became the classroom in this creative endeavour.

Contemporary views of our physical world suggest that there exist but two ultimate natural resources: matter and energy. Matter appears in various forms to provide a range of substances; in the language of an elementary school game, it is either animal, vegetable or mineral. Energy, like matter, may also exist in various forms -- in the thundering motion of ocean waves, in the action of a heart valve, and in the vibrating motion of a guitar string.

It is common to observe manifestations of both matter and energy. A piece of burning wood provides heat energy; a quantity of gasoline enables automobile transport; a photon of radiation can affect the growth of a tumor. Indeed, in our modern view of the universe, matter and energy are equivalent.

1.2 Matter-Energy Connections

The most pronounced consequences of our physical world are associated with the dynamical processes involving matter and energy; it is the inherent transformations that have long constituted a subject of abiding interest. For this purpose, descriptions of a state and of a process need to be established.

When matter in the form of water drops unimpeded through a vertical distance, it acquires kinetic energy. This energy may be used to turn a water turbine which, when connected to a generator, produces electrical energy E_e given by

$$E_e = \eta \, mgh \, . \tag{1.1}$$

Here m is the relevant mass of water, g is the suitably averaged local acceleration due to gravity, h is the vertical distance of fall, and η is the efficiency of converting gravitationally sustained energy into electrical form.

We may characterize the production of electrical energy by this gravitational process as a transformation of matter m from an initial vertical coordinate y_i to a final coordinate y_f

$$mgy_i \rightarrow mgy_f \, . \tag{1.2}$$

Note the various factors which contribute to this gravitational-to-electrical transformation of energy: m is a natural resource, g is a local geophysical parameter, h is determined by circumstances of geography and hydrology, and η is governed by the material limitations of the energy conversion device.

A similar characterization may be associated with the common combustion process of, say, methane, CH_4, in the presence of molecular oxygen O_2, under suitable conditions of temperature and pressure:

$$CH_4 + 2O_2 \rightarrow 2H_2O + CO_2 . \tag{1.3}$$

Both CH_4 and O_2 represent a natural resource of matter and the imposition of a specific pressure involves a man-made device; the associated temperature manifests itself in a particular state of molecular agitation, creating conditions for the reaction to proceed. We may therefore more correctly assert that an initial ignition energy, E_i, needs to be supplied to CH_4 and to $2O_2$ resulting in the rearrangement of atoms yielding $2H_2O$ and CO_2 possessing kinetic energy E_f. The transformation process may therefore be represented as

$$E_i + CH_4 + 2O_2 \rightarrow 2H_2O + CO_2 + E_f . \tag{1.4}$$

The net energy generated in such a transformation is therefore $(E_f - E_i)$ and may in general be exothermic or endothermic, depending upon whether E_f is greater or less than the required ignition energy, E_i.

1.3 Mass-Energy Equivalence

The preceding discussion suggests a connection between matter and energy at one level of common experience. However, the most dramatic manifestations of such a linkage occur at the level of nuclear transformations.

A detailed analysis of colliding particles when viewed from two inertial coordinate systems moving relative to each other suggests that the mass of an object depends upon its speed. For the case of an object of rest mass m_r, it is found that when it is moving at speed v, its mass m_v is related to m_r by

$$m_v = \frac{m_r}{\sqrt{1 - v^2/c^2}} . \tag{1.5}$$

where c is the speed of light in free space -- a constant independent of the relative speed of the observer*. For moving objects of common experience -- aircraft, tennis balls, missiles -- we have $v << c$, and hence, $m_v \simeq m_r$. However, subatomic particles can attain speeds approaching that of light and then $m_v > m_r$. For example, at a speed 10% that of light, an electron's mass increases by about 1%, while at 0.9 c its mass has more than doubled. According to Eq. (1.5), the mass of an object approaches infinity as $v \rightarrow c$ and therefore would require an infinite energy supply to accelerate it to that speed; hence, c represents the upper limit of particle speed in free space, a result confirmed by actual experiment with nuclear accelerators.

An important consequence results from the binomial expansion of the radical of Eq. (1.5):

$$m_r = m_v[1 - v^2/c^2]^{1/2}$$

$$= m_v\left[1 - \frac{1}{2}\frac{v^2}{c^2} + \frac{\left(\frac{1}{2}\right)\left(\frac{1}{2}-1\right)}{2!}\left(\frac{v^2}{c^2}\right)^2 - \frac{\left(\frac{1}{2}\right)\left(\frac{1}{2}-1\right)\left(\frac{1}{2}-2\right)}{3!}\left(\frac{v^2}{c^2}\right)^3 + \ldots\right]. \quad (1.6)$$

Multiplying by c^2 and rearranging, we obtain

$$m_v c^2 = m_r c^2 + \frac{1}{2} m_v v^2 + \frac{1}{8} m_v\left(\frac{v^4}{c^2}\right) + \frac{3}{48} m_v\left(\frac{v^6}{c^4}\right) + \ldots , \quad (1.7)$$

and may now attach the following important interpretation. First, each term possesses units of energy. Second, the first term on the right possesses an intrinsic property of the particle at rest; all other terms on the right are non-zero only if the particle is in motion and hence represent the energy of the particle attributable to motion, that is kinetic energy. We therefore write an expression for the rest mass energy

---

* See Appendix A for numerical value of this and other constants.

$$E_r = m_r c^2 , \tag{1.8}$$

and the remaining terms represent the kinetic energy

$$E_k = \frac{1}{2} m_v v^2 + \frac{1}{8} m_v \left(\frac{v^4}{c^2}\right) + \frac{3}{48} m_v \left(\frac{v^6}{c^4}\right) + \ldots ,$$

$$= \frac{1}{2} m_v v^2 \left\{ 1 + \frac{1}{4}\left(\frac{v}{c}\right)^2 + \frac{3}{24}\left(\frac{v}{c}\right)^4 + \ldots \right\} , \tag{1.9}$$

The total energy of a particle is therefore

$$E_t = E_r + E_k , \tag{1.10}$$

or

$$m_v c^2 = m_r c^2 + E_k . \tag{1.11}$$

Three important conclusions thus emerge: One is that an object, even when stationary, contains energy called rest mass energy, Eq. (1.8). Then, at low speed $v << c$, only the leading term of the kinetic energy component is important so that

$$E_k \simeq \frac{1}{2} m_v v^2 \simeq \frac{1}{2} m_r v^2 , \tag{1.12}$$

and corresponds to the classical expression of kinetic energy of a moving particle. Finally, it is the mass of the object and not the form of matter which is important.

1.4 Mass-Energy Conservation

The recognition that energy is associated with matter-in-motion is quite evident because it conforms to ordinary experience; common examples of such observations are moving trains, vibrating strings, spinning tops and innumerable others. Equation (1.10), however, asserts much more: the total energy of any piece of matter consists of two components, one associated with its motion and one independent of its motion. It is this

Discussion / Analysis 1.1

Computation of rest-mass energies and comparison with other energy quantities.

$$E_r = m_r c^2$$

- c — speed of light (m/s), Table A.2
- m_r — rest mass of object (kg)
- E_r — rest mass energy "contained" in object (J)

For an electron:

$$E_r = (9.1 \times 10^{-31}\,\text{kg})(3 \times 10^8\,\text{m/s})^2 = \boxed{8.2 \times 10^{-14}\,\text{J}}$$

(9.1×10^{-31} kg: Table A.1; also 0.511 MeV)

For a pin head:

$$E_r = (10^{-4}\,\text{kg})(3 \times 10^8\,\text{m/s})^2 = \boxed{9 \times 10^{12}\,\text{J}}$$

(10^{-4} kg: estimate)

Compare (consult energy handbook):

1 barrel of oil: $\boxed{\sim 10^{10}\,\text{J}}$

Avg. global per capita energy consumption in one year: $\boxed{\sim 10^{11}\,\text{J/y}}$

- ~ 10 barrels of oil per year
- ~ $\frac{1}{100}$ of a pin head per year

To think about:

How might the energy "locked" up in matter be extracted?

stationary rest mass energy component which is so intriguing because this statement claims that energy resides intrinsically in all matter. We consider this important concept further.

Equations (1.3) and (1.4) indicate that a rearrangement of atoms in their molecular combination has taken place during the reaction. That is, for reactions in general, we should write

$$a + b \rightarrow \begin{pmatrix} \text{rearrangement} \\ \text{of components} \end{pmatrix} \rightarrow d + e \,, \tag{1.13}$$

though the details of the "rearrangement" are beyond the scope of this text.

Similar rearrangements occur at a more spectacular level if the interacting entities "a" and "b" are nuclei. It is known that nuclides are composed of nucleons and that during a nuclear reaction new nucleon groupings occur. Consider then a nuclear particle "a" possessing a rest mass $m_{r,a}$ and kinetic energy $E_{k,a}$ on a collision course with nuclear particle "b" of rest mass energy $m_{r,b}$ and kinetic energy $E_{k,b}$, Fig. 1.1.

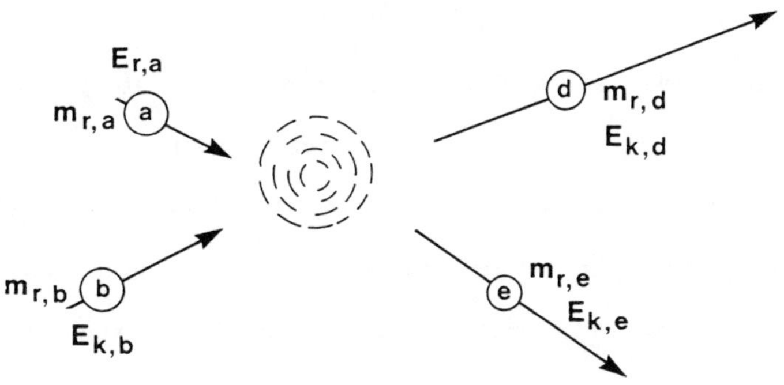

Fig. 1.1: Nuclear particles "a" and "b" on a collision course. As a result of the collision, new particles "d" and "e" appear.

The combined mass-energy conservation principle which now applies is that the initial total energy $E_{t,i}$ and final total energy $E_{t,f}$ are equal:

$$E_{t,i} = E_{t,f}. \tag{1.14}$$

Then, showing specifically the rest mass energy components and kinetic energy component, we have

$$E_{r,i} + E_{k,i} = E_{r,f} + E_{k,f}. \tag{1.15}$$

Substituting for the rest mass-energies for particles a, b, d and e, and also specifically identifying the kinetic energies,gives

$$[(m_{r,a} c^2 + E_{k,a}) + (m_{r,b} c^2 + E_{k,b})] = [(m_{r,d} c^2 + E_{k,d}) + (m_{r,e} c^2 + E_{k,e})] . \tag{1.16}$$

Next, we rearrange these terms to yield

$$(E_{k,d} + E_{k,e}) - (E_{k,a} + E_{k,b}) = [(m_{r,a} + m_{r,b}) - (m_{r,d} + m_{r,e})] c^2 , \tag{1.17}$$

or, more succinctly,

$$E_{k,f} - E_{k,i} = (m_{r,i} - m_{r,f}) c^2 . \tag{1.18}$$

Alternatively

$$\Delta E_k = - (\Delta m_r) c^2 , \tag{1.19}$$

where here and elsewhere in this text, the delta symbol $\Delta(\ )$ is always used to represent {"$(\ )_{after} - (\ )_{before}$"}. Equation (1.19) represents a most remarkable result: if, following a nuclear interaction, a net decrease in the total rest mass has occurred, that is $(\Delta m_r) < 0$, then a gain in kinetic energy of the resultant reaction products has taken place. The converse also holds true. We suggest this important relationship in Fig. 1.2:

Two questions now arise: are there readily sustainable reactions for which $\Delta m_r < 0$ and how may the increase in net kinetic energy be extracted into useful form? In answer to the first part, it may be asserted that there are indeed readily sustainable

Fig. 1.2: Depiction of equivalence between mass and energy as a consequence of a nuclear transformation.

nuclear reactions which are highly exoergic. Concerning the recovery of kinetic energy of the reaction products, recall that fast-moving nuclear particles will interact with other atoms by collisional processes and thereby heat the material in which the reactions occur. Thus, energy which initially resides in the motion of nuclear particles may be converted into heat and subsequently further converted into other forms of energy.

The relative magnitude of energy released for gravitational, chemical and nuclear processes on a per-event basis is suggested in Table 1.1. Note that nuclear reactions are order-of-magnitude more exoergic than others.

1.5 Units and Quantities

The universal standard of mass is the kilogram defined by an internationally accepted reference mass. Energy is a derived quantity and is expressible in terms of the base units of kilogram (kg), meter (m), and second (s). The SI unit for energy is the joule, J, defined in dimensional form by

$$1\,\mathrm{J} = 1\,\mathrm{kg\,m^2\,s^{-2}}\ . \qquad (1.20)$$

Table 1.1

Comparison of energies associated with various processes

	Transformation	Energy release per event (joules)	Normalized ratio
I.	Gravitational H_2O molecule dropping 100 m	$\sim 10^{-23}$	1
II.	Chemical Typical hydro-carbon combustion	$\sim 10^{-18}$	$\sim 10^{5}$
III.	Nuclear		
	De-excitation of a nucleus	$\sim 3 \times 10^{-13}$	$\sim 10^{10}$
	Fusion of hydrogen isotopes	$\sim 4 \times 10^{-12}$	$\sim 10^{11}$
	Fission of heavy element	$\sim 5 \times 10^{-11}$	$\sim 10^{12}$

Other derived energy units have become common in specialized application. The most frequently used and their equivalents to the joule are the following:

$$\begin{aligned} 1\ \mathrm{J} &= 6.24 \times 10^{18}\ \mathrm{eV}\ \text{(nuclear usage)} \\ &= 2.78 \times 10^{-7}\ \mathrm{kWh}\ \text{(electrical usage)}\ . \end{aligned} \tag{1.21}$$

Power is the rate of energy transformation and, in the language of the differential calculus, is given by

$$P = \frac{dE}{dt} , \tag{1.22}$$

with its unit being the watt, W, defined by 1 W = 1 J/s. Power may therefore be interpreted as an"energy current". Further, according to Eq. (1.22), energy can be expressed as the time integral of power

$$E(t) = \int_0^t P(t')\,dt' , \tag{1.23}$$

with E(0) = 0. Thus, while power is an "energy current", energy is the "area under the power curve". If power is a constant during a time period τ, then we have simply

$$E = P\tau, \qquad P = E/\tau. \tag{1.24}$$

A range of energies and powers are in common use; we display some of these in Fig. 1.3.

1.6 Energy, Power and Reactions

The release of energy resulting from a rearrangement of nuclear structure is a fundamental theme in this text. Such a rearrangement may occur as a consequence of two-body collisions, Sec. 1.4, or it may occur spontaneously in unstable nuclides. If a particular type of such exoergic nuclear events occurs in a reaction domain at the rate R(t) events per unit time and if the energy concurrently released is Q joules per event, e.g. Eq. (1.19), then the instantaneous nuclear power thus generated is

$$P(t) = R(t)\,Q. \tag{1.25}$$

The total energy released up to time t is therefore

$$E(t) = Q \int_0^t R(t')\,dt'. \tag{1.26}$$

Note that Q is a constant with respect to time.

If several distinguishable nuclear events occur concurrently in the reaction domain, then evidently

$$P(t) = \sum_i R_i(t)\,Q_i, \tag{1.27}$$

and

$$E(t) = \sum_i \left\{ Q_i \int_0^t R_i(t')\,dt' \right\}. \tag{1.28}$$

Energy, power, and reactions are thus closely related.

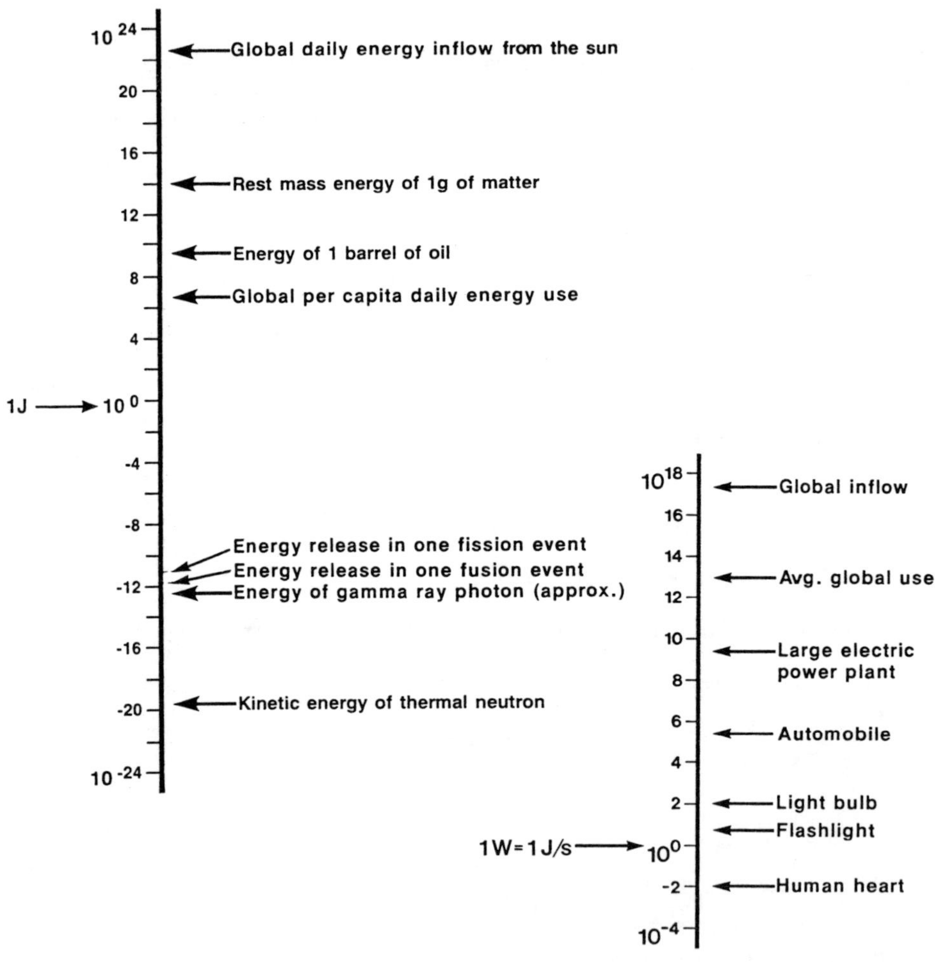

Fig. 1.3: Graphical depiction of energies and powers associated with various processes.

Discussion / Analysis 1.2

The sun's rays originate in the "nuclear burning" of hydrogen. How much mass is thus burned in a second?

SUN

$-\frac{\Delta m}{\Delta t}$

$$P_{sun} = \frac{\Delta E}{\Delta t} \sim 4 \times 10^{26}\ J/S$$

astronomy handbook

Know

$$\Delta E = (-\Delta m)c^2$$

mass burned to yield ΔE

energy resulting from burning Δm.

During time Δt

$$\frac{\Delta E}{\Delta t} = \left(-\frac{\Delta m}{\Delta t}\right)c^2$$

mass burning rate $\dot{m}$ (kg/s)

power, P_{sun} (J/s)

$$\therefore -\frac{\Delta m}{\Delta t} = \frac{\Delta E/\Delta t}{c^2} = \frac{P_{sun}}{c^2}$$

$$= \frac{4 \times 10^{26}\ J/S}{(3 \times 10^{8}\ m/s)^2} = \boxed{4.4 \times 10^{9}\ kg/s}$$

To think about:

Knowing that the present solar mass is $\sim 2 \times 10^{30}$ kg, how long until 10% of the sun's mass is burned?

Problems

1.1 Distinguish between matter and mass.

1.2 Confirm the correctness of the gravitational energy transformation listed in Table 1.1.

1.3 Determine the power required to operate a typical household refrigerator. Also, compute its annual energy requirements.

1.4 Use Eq. (1.8) to calculate the total rest mass energy content of a coin. Then, using the local price for electrical energy, determine the "$E = mc^2$" energy value of the coin.

1.5 Electrons in a high energy accelerator may attain a total energy of 5 GeV. What is the mass and speed of the electrons at this energy?

1.6 Compute your energy cost per joule for each of the following:

a. electricity

b. gasoline

c. coal

d. natural gas.

1.7 Use Table 1.1 and Eq. (1.25) to determine the number of fission reactions occurring each second in a 100 MW_t fission reactor.

CHAPTER II

NUCLEAR STRUCTURE AND PROPERTIES

Nuclides are characterized by properties quite different from objects commonly encountered. These can be described by models useful in providing a conceptual understanding of the nucleus and nuclear phenomena.

2.1 Nuclear Composition

An atom may be viewed as an object consisting of an inner high-density core surrounded by electrons orbiting in discrete energy states. The nuclear core consists of positively charged particles called protons and neutral particles called neutrons; these particles are collectively known as nucleons. The rest masses and charges of the electrons, protons and neutrons are known to high accuracy, Table 2.1.

The "radius" of the nucleus, R, containing A nucleons (protons and neutrons), each of "radius" r, can be estimated by volume conservation

$$A\left(\frac{4}{3}\pi r^3\right) \simeq \frac{4}{3}\pi R^3, \tag{2.1}$$

giving therefore an $A^{1/3}$ dependence

$$R \simeq r A^{1/3}, \tag{2.2}$$

with $r \sim 10^{-15}$ m; see Table 2.1 for other dimensions.

A nuclide is a nuclear species uniquely characterized by its number of protons, Z, and its number of neutrons, N. It is identified by the following three equivalent notations:

$$^{Z+N}_{Z}X = ^{A}_{Z}X = ^{A}X . \tag{2.3}$$

Table 2.1

Some nuclear and atomic properties

Object	Rest Mass (kg)	Atomic Mass* (u)	Charge (C)	"Radius" (m)
Electron	9.10956×10^{-31}	5.48584×10^{-4}	$(-)1.60219 \times 10^{-19}$	
Proton	1.67261×10^{-27}	1.00726	$(+)1.60219 \times 10^{-19}$	$\sim 10^{-15}$
Neutron	1.67492×10^{-27}	1.00865	0	$\sim 10^{-15}$
Nucleus	$\sim 1.67 \times 10^{-27}$ A	–	$(+)1.6 \times 10^{-19} \times Z$	$\sim 10^{-15} A^{1/3}$
Atom	$\sim 1.67 \times 10^{-27}$ A	–	0	$\sim 10^{-10}$

* The "atomic mass unit" (u) is defined as 1/12 of the carbon-12 atom and is related to the kilogram mass unit by $1\ u = 1.66056 \times 10^{-27}$ kg.

Here the nucleon number is A = Z + N and the symbol X is the elemental name defined by the number of protons in the nucleus; listing both Z and X is therefore redundant and the Z subscript is frequently not listed.

With Z and N as integer identifiers of a nucleus, we may represent all known nuclides on a Cartesian coordinate system as suggested in Fig. 2.1.

In Fig. 2.2, we suggest the identification of an arbitrary nucleus (N, Z) and its immediate neighbours. These differ by an integral number of protons or neutrons or both, suggesting that the dream of the medieval alchemists could be realized by the simple addition or removal of nucleons. Indeed, nuclear research facilities and all nuclear power

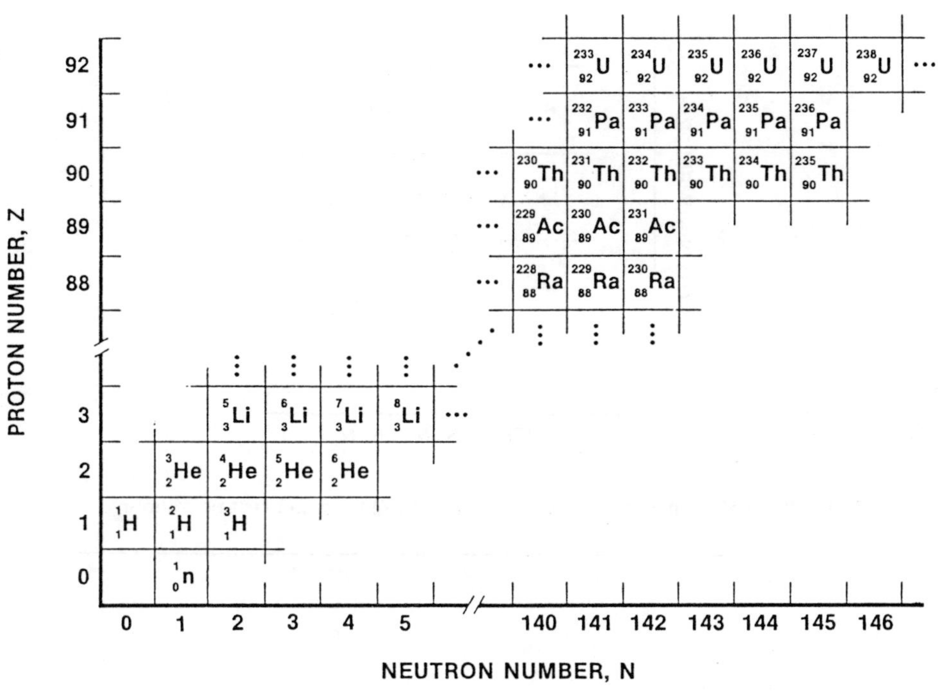

Fig. 2.1: Depiction of some nuclides on a N-Z Cartesian grid. Such a representation is frequently called a Segrè Chart or Chart of the Nuclides; the entire chart, together with additional descriptive and numerical information, is found as an insert of the back-cover page.

reactors are designed for the transformation of selected nuclides including also processes other than those implied in Fig. 2.2.

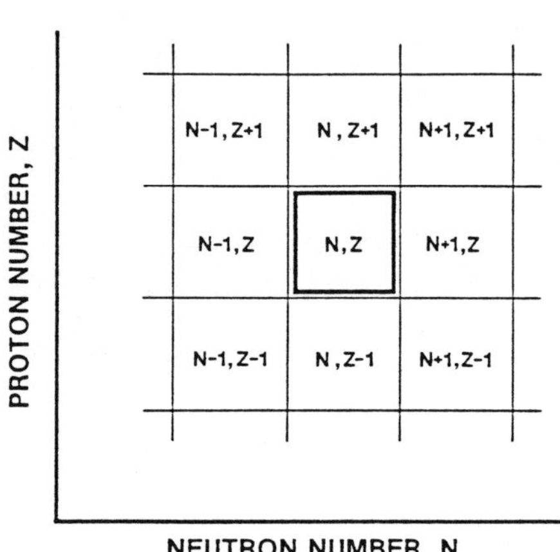

Fig. 2.2: Relationship of a given nuclide (N, Z) to its nearest neighbours.

2.2 Binding Energy

Accurate measurements of nuclear masses exist. The carbon-12 nucleus (Z = 6, N = 6) is known to possess a rest mass of

$$m_c = 19.9209 \times 10^{-27}\,\text{kg}. \tag{2.4}$$

Now, according to Table 2.1, the total rest mass of 6 isolated protons and 6 neutrons is

$$\begin{aligned} 6\,m_p &= 10.0357 \times 10^{-27}\,\text{kg} \\ 6\,m_n &= 10.0495 \times 10^{-27}\,\text{kg} \\ \hline &\ 20.0852 \times 10^{-27}\,\text{kg}, \end{aligned} \tag{2.5}$$

revealing therefore that, at the nuclear level of physical reality, the whole is not the sum of its parts.

Discussion / Analysis 2.1

Computation of mass density of a nucleus and comparison to other densities.

Def. $\rho = \dfrac{\text{mass}}{\text{volume}}$

$$= \frac{m\,A}{\frac{4}{3}\pi R^3}$$

m — mass of nucleon

A — number of nucleons in nucleus

R — radius of nucleus, $R = r A^{1/3}$ (see 2.1)

$$= \frac{m\,A}{\frac{4}{3}\pi (r A^{1/3})^3}$$

$$= \frac{3\,m}{4\pi r^3} \simeq \frac{3 \times 1.67 \times 10^{-27}\,\text{kg}}{4\pi \times (10^{-15}\,\text{m})^3}$$

$$= 4 \times 10^{14}\,\frac{\text{kg}}{\text{m}^3} = \boxed{4 \times 10^{11}\,\frac{\text{g}}{\text{cm}^3}}$$

Compare (reference texts)

$$\rho_{H_2O} = \boxed{1\,\frac{\text{g}}{\text{cm}^3}}; \qquad \rho_{Pb} = \boxed{11.3\,\frac{\text{g}}{\text{cm}^3}}$$

To think about:

What volume would a typical locomotive occupy if all the electrons from its constituent atoms were removed and the nuclides compressed into a "nuclear matter" sphere?

A useful characterization of this important observation can be formulated if we consider the process of forming an arbitrary nuclide of A nucleons from a union of Z protons and N neutrons. While the details may indeed be very complex, the overall process may be represented by

$$Zm_p + Nm_n \rightarrow m_A , \tag{2.6}$$

where rest masses for the protons, neutrons and the entire nucleus are assumed. Using again the conventional definition of an increment for a process $\Delta(\ ) = (\ )_{after} - (\ )_{before}$ we have for Eq. (2.6)

$$\Delta m = m_A - (Zm_p + Nm_n) . \tag{2.7}$$

This mass difference is negative for all known nuclides; that is the rest mass of the parts is greater than the rest mass of the whole! The important question now is, what does this difference imply? We provide here two comments.

First, a rational interpretation is that the mass decrement Eq. (2.7) corresponds, by the mass-energy equivalance relations of Chapt. I, to the energy which holds the nucleus together. That is, in order to sustain a tight configuration of protons and neutrons, energy is required to oppose the long-range electrostatic repulsion of the positively charged protons. We call this the nuclear "binding" energy E_b -- characterized by short-range effects -- and may compute it from

$$\begin{aligned} E_b &= -[m_A - (Zm_p + Nm_n)]\, c^2 \\ &= -(\Delta m)\, c^2 . \end{aligned} \tag{2.8}$$

Dividing this quantity by the nucleon number A then yields an average energy for "binding" a nucleon in a nucleus:

$$\overline{E}_b = \frac{E_b}{A} . \tag{2.9}$$

We display this quantity for all nuclides in Fig. 2.3 below. The interesting result here is that for very light nuclides, each additional nucleon becomes increasingly more tightly bound but that this trend eventually decreases.

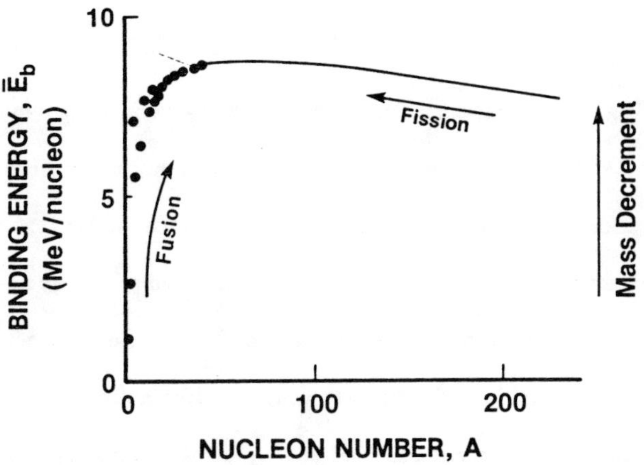

Fig. 2.3: The Curve of Binding Energy.

Second, another useful interpretation follows from the recognition that all nuclides appear somewhere on the curve of binding energy and that each possesses a different amount of energy locked-up, so to speak, in holding it together. Supposing it were possible to combine two light nuclides to form one heavier or to break-up a heavy nucleus into some lighter ones. While such a process represents considerable nuclear reorganization each of the reaction products would still be found somewhere on the Curve of Binding Energy. Then, if little energy was required in bringing these transformations about, the associated mass decrement may lead to a significant energy gain in the form of increased motion of the reaction products, Sec. 1.4. Such reaction processes are indeed known: the fissioning of some heavy nuclei by thermal neutron absorption is an existing and available technology for energy production while the fusioning of light nuclei is possible if efficient means of heating them and then confining them can be found. We also suggest these two kinds of transformations in Fig. 2.3.

2.3 Energy Levels

According to atomic theory, the electron orbits surrounding a nucleus exist in well defined energy states. When an electron "drops" from a higher shell to one nearer to the core, the difference in its energy state appears in the form of X-rays. Conversely, if an electron is to be elevated to a higher level, a precise amount of energy has to be supplied.

A similar characterization may be applied to a nucleus. There exists a ground state and various elevated levels which are unique -- like a distinct fingerprint -- to each nucleus. One nuclear property is that the energy level structure becomes more complex with increasing atomic mass number. We show such nuclear levels for two nuclides in Fig. 2.4.

A nucleus may be excited to a particular energy level upon the absorption of a specific amount of energy introduced either by electromagnetic radiation or by nuclear collisional effects. (A third possibility is that it may appear in an excited state as the result of nuclear disintegration of its parent, to be discussed in the next chapter.) In either case, the nucleus generally de-excites within a very short time by the emission of one or several gamma rays; the energies of these emitted gammas correspond precisely to the allowed differences in energy states of the nucleus.

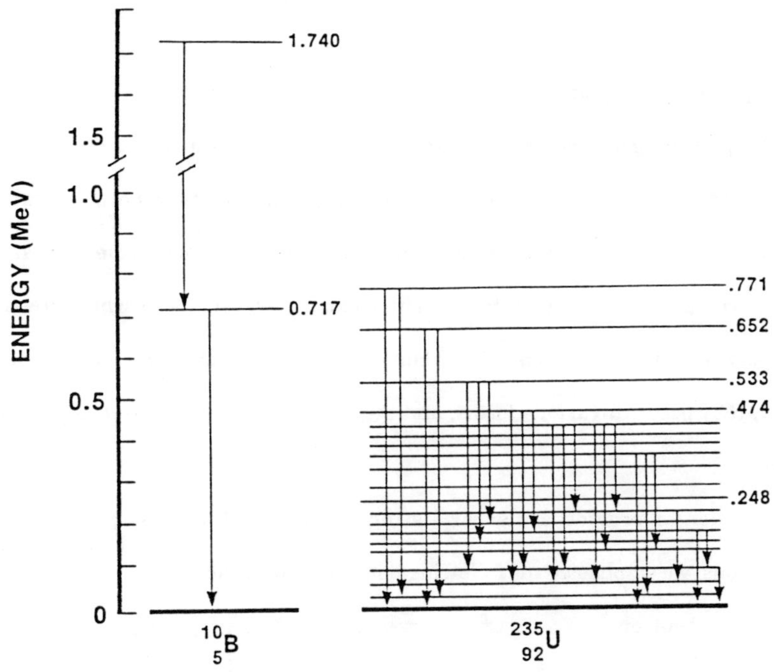

Fig. 2.4: Illustration of nuclear energy levels for a light (boron-10) and a heavy (uranium-235) nucleus; not all levels for ^{235}U are shown. The arrows indicate permissible transitions with the attendant release of a gamma ray photon.

A close connection between the gamma ray photon energy E_γ and its frequency ν_γ is given by the relationship

$$E_\gamma = h\,\nu_\gamma\ , \tag{2.10}$$

where h is Planck's constant. Its speed is that of light in the medium c and thus defines a gamma wavelength λ_γ given by

$$\lambda_\gamma\,\nu_\gamma = c\ . \tag{2.11}$$

Excited nuclear states whose mean life-time is longer than a specified length of time ($\sim 10^{-6}$ s) are frequently called isomer or metastable states.

2.4 Nuclear Disintegration

The preceding discussion of dynamic nuclear processes involved the absorption of energy and the emission of gamma rays. An important feature here is that in the processes of nuclear deexcitation -- that is, the attainment of a more stable nuclear state -- the nuclear composition was not altered. There exist, however, states which are unstable with respect to processes which affect their composition; such phenomena occur spontaneously, they occur in various forms, and they are unaffected by external macroscopic conditions.

In order to provide a clearer listing of some of the most common nuclear disintegrations, we emphasize the following notation to be used:

n: neutron

p: proton

α: alpha particle (helium-4 nucleus)

β^-: beta particle (an electron originating in the nucleus)

β^+: positron (a positively charged electron originating in the nucleus)

P_i: fission product of unspecified composition

The spontaneous ejection of a neutron may, according to the schema of Sec. 2.1, be represented by

$$(N, Z) \rightarrow (N-1, Z) + n\,, \tag{2.12a}$$

of which the following is a specific example:

$$^{87}_{35}Br \rightarrow {}^{86}_{35}Br + n\,. \tag{2.12b}$$

Notice that the elemental name of the nuclides has not changed though the nucleon number did decrease by an integer.

The ejection of an alpha particle involves 2 neutrons and 2 protons, e.g.:

$$(N, Z) \rightarrow (N-2, Z-2) + \alpha \, . \tag{2.13a}$$

Such a process occurs frequently with heavy elements; the following is an example:

$$^{236}_{92}U \rightarrow \, ^{232}_{90}Th + \alpha \, . \tag{2.13b}$$

Here, a new element has emerged from this disintegration.

The ejection of negative and positive electrons, i.e. β^- and β^+ respectively, involves an additional process occurring in the nucleus. For β^- ejection we have

$$n \rightarrow p + \beta^- , \tag{2.14}$$

so that the nucleon number A does not change though the neutron number N decreases and the proton number Z increases:

$$(N, Z) \rightarrow (N-1, Z+1) + \beta^- \, . \tag{2.15a}$$

The following is such an example

$$^{28}_{13}Al \rightarrow \, ^{28}_{14}Si + \beta^- \, . \tag{2.15b}$$

Then, positron ejection in general involves

$$p \rightarrow n + \beta^+ , \tag{2.16}$$

so that,

$$(N, Z) \rightarrow (N+1, Z-1) + \beta^+ \, . \tag{2.17}$$

There exist nuclear disintegrations involving both β^- and β^+ ejection occurring, however, with unique branching probabilities of occurrence. An example is the following:

$$^{74}As \; \underset{0.68}{\overset{0.32}{\longrightarrow}} \; \begin{matrix} \nearrow \; ^{74}Se + \beta^- \\ \searrow \; ^{74}Ge + \beta^+ \end{matrix} \, . \tag{2.18}$$

All the above nuclear disintegration processes are frequently described as spontaneous nuclear transmutations or nuclear decays; their graphical depiction is suggested in Fig. 2.5 below at an arbitrary location of the Chart of the Nuclides.

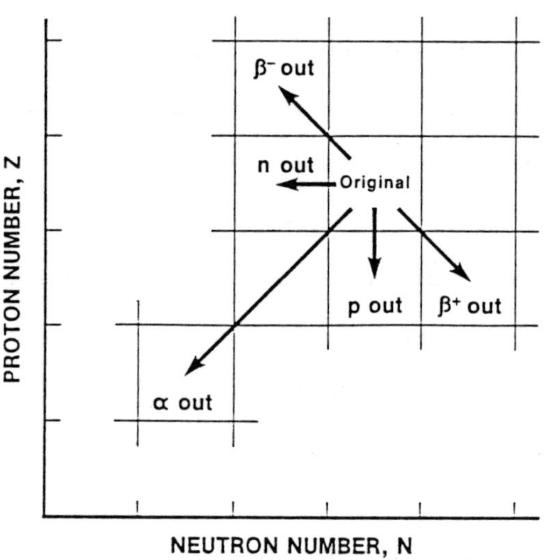

Fig. 2.5: Depiction of nuclear transmutation by the emission of various particles.

Occasionally, however, even a more comprehensive nuclear rearrangement may occur. One such spontaneous disintegration may be written as

$$^{238}U \rightarrow \nu n + P_1 + P_2, \qquad (2.19)$$

with the reaction products P_1 and P_2, as well as the number of emitted neutrons ν, as statistical variables subject to the conservation of nucleons. Because the products of this disintegration are indistinguishable from a neutron-induced fission event, it is called spontaneous fission.

Very often the reaction products are themselves radioactive and generate further disintegration products and/or gamma rays. In some cases, well defined nuclear chains can be identified as suggested in the following:

$$
\begin{array}{l}
{}^{232}Th \rightarrow {}^{228}Ra + \alpha \\
\quad\swarrow \\
{}^{228}Ra \rightarrow {}^{228}Ac + \beta^{+} \\
\quad\swarrow \\
{}^{228}Ac \rightarrow \ldots \\
\ldots
\end{array}
\tag{2.20}
$$

Each of the nuclear decay processes enumerated above involves a mass decrement and hence can be associated with a mass-energy transformation, Eq. (1.18). In view of the several reaction products, and based on our discussion of Sec. 1.4,we write

$$E_{k,f} - E_{k,i} = -\left[\sum_{f} (m_{r,f}) - m_{r,i}\right] c^2 . \tag{2.21}$$

Then, in general $E_{k,i} << E_{k,f}$ so that the energy associated with the mass decrement $-(\Delta m)c^2$ appears predominantly in the form of kinetic energy of the decay products $E_{k,f}$.

2.5 Collisional Phenomena

Collisions between nuclides and nucleons may be depicted in some cases as

$$a + b \rightarrow d \ , \tag{2.22a}$$

and in other cases as

$$a + b \rightarrow e + f + \ldots \tag{2.22b}$$

Here, Eq. (2.22a) represents a capture or compound nucleus formation process; the second process, Eq. (2.22b), involves a very rapid intermediate unstable dynamic process followed by disintegration. Hence, there may be an occasion when it is desirable to write

Discussion / Analysis 2.2:

Tritium is a beta emitter:

$$^{3}H \longrightarrow {}^{3}He + \beta^{-} + \bar{\nu}$$

Obtain an expression for the maximum kinetic energy of β^{-}.

Before decay (n, n, p) — After decay (n, p, p), $\bar{\nu}$, β^{-}

Mass - energy relation

$$\Delta E = (-\Delta m_r)c^2$$

$$\left[E_{k,after} - E_{k,before}\right] = -\left[m_{r,after} - m_{r,before}\right]c^2$$

$$\left[(E_{k,He} + E_{k,\beta} + E_{k,\nu}) - E_{k,H}\right] = -\left[(m_{r,He} + m_{r,\beta} + m_{r,\nu}) - m_{r,H}\right]c^2$$

$E_{k,He} \sim 0$, $E_{k,\beta}$ max, $E_{k,\nu} \sim 0$ (case of interest); $E_{k,H} = 0$; $m_{r,\beta} \ll m_{r,He}$; $m_{r,\nu} \to 0$

$$\therefore \quad \boxed{(E_{k,\beta})_{max} = \left[m_{r,H} - m_{r,He}\right]c^2}$$

To think about:

Note that $m_{r,H} > m_{r,He}$ even though $Z_H < Z_{He}$!

$$a + b \rightarrow (ab)^* \rightarrow e + f + \cdots \tag{2.23}$$

where $(ab)^*$ represents such a very short-lived unstable nuclide. Note that here the first arrow represents a nuclear formation process and the second arrow a nuclear disintegration or decay process.

The following are examples of the more important nuclear collisional processes of particular importance in nuclear engineering:

Neutron capture:	$^{112}Cd + n \rightarrow {}^{113}Cd$,	(2.24a)
Neutron-induced fission:	$^{235}U + n \rightarrow \nu n + P_1 + P_2$,	(2.24b)
Fusion of hydrogen isotopes:	$^{2}H + {}^{3}H \rightarrow n + \alpha$,	(2.24c)
Breeding of fusile fuel:	$^{6}Li + n \rightarrow {}^{3}H + \alpha$,	(2.24d)
Breeding of fissile fuel:	$^{232}Th + n \rightarrow {}^{233}Th$	
	$^{233}Th \rightarrow {}^{233}Pa + \beta^-$	(2.24e)
	$^{233}Pa \rightarrow {}^{233}U + \beta^-$.	

Note, the sequential "chaining" reactions of Eq. (2.24e).

The first four equations here, Eqs. (2.24a) to (2.24d) are written in a form commonly used in the literature and follow the notational pattern of Eq. (2.22). Based on Eq. (2.23), these same equations would appear as follows:

$$^{112}Cd + n \rightarrow ({}^{113}Cd)^* \rightarrow {}^{113}Cd + \gamma , \tag{2.25a}$$

$$^{235}U + n \rightarrow ({}^{236}U)^* \rightarrow P_1 + P_2 + \nu n , \tag{2.25b}$$

$$^{2}H + {}^{3}H \rightarrow ({}^{5}He)^* \rightarrow n + \alpha , \tag{2.25c}$$

$$^{6}Li + n \rightarrow ({}^{7}Li)^* \rightarrow {}^{3}H + \alpha . \tag{2.25d}$$

In the first equation, Eq. (2.25a), $({}^{113}Cd)^*$ is simply the Cd-113 nucleus in an excited state and it promptly decays to its ground state by gamma emission.

An interesting and also very important event is associated with $({}^{236}U)^*$ of Eq. (2.25b). There exists a fixed probability that rather then undergoing fission it may, like

$(^{113}Cd)^*$, simply emit a gamma ray. These two competing reaction channels may be represented by the following:

$$^{235}U + n \rightarrow (^{236}U)^* \begin{array}{c} \xrightarrow{x} \nearrow \ ^{236}U + \gamma \\ \xrightarrow[1-x]{} \searrow \ P_1 + P_2 + \nu n \end{array} \tag{2.26}$$

Typically the capture probability $x = 0.15$ so that neutron absorption in ^{235}U leads to fission only 85% of the time. Further, P_1 and P_2 are invariably unstable and decay by either particle or gamma-ray emission, or both.

The case of $(^5He)^*$ in Eq. (2.25c) is interesting because its life-time is so short that it cannot be detected; a similar situation applies to $(^7Li)^*$ though it does exist independently as a stable nucleus.

While Eq. (2.21) refers to the energy release associated with nuclear disintegration, we can similarly compute the mass decrement associated with nuclear collisions by writing

$$E_{k,f} - E_{k,i} = -\left[\sum_f (m_{r,f}) - \sum_i (m_{r,i})\right]c^2. \tag{2.27}$$

Here $m_{r,f}$ and $m_{r,i}$ are particular final and initial particle rest masses.

Equation (2.27) is of such importance and is used so frequently that it is commonly written as

$$Q = -\left[\sum_i (m_{r,i}) - \sum_f (m_{r,f})\right]c^2, \tag{2.28}$$

and thus defines the Q-value of a nuclear reaction.

2.6 The Early Universe

The primeval formation of elements and associated nuclear energy releases is conceived as having been initiated by the "burning" of hydrogen during the gravitational collapse of a stellar proton gas. The primary initiating fusion process is conceived to have been

$$^{1}H + {^{1}H} \rightarrow {^{2}H} + \beta^{+} + \bar{\nu} \,, \tag{2.29}$$

with ν a neutrino. The deuteron thus formed reacts with a background proton to form helium-3:

$$^{1}H + {^{2}H} \rightarrow {^{3}He} \,. \tag{2.30}$$

Subsequently, the helium-3 reaction product fuses with another helium-3 nucleus to yield an alpha particle and two protons,

$$^{3}He + {^{3}He} \rightarrow {^{4}He} + 2\,{^{1}H} \,. \tag{2.31}$$

The next heavier element produced is beryllium, by the process

$$^{3}He + {^{4}He} \rightarrow {^{7}Be} \,, \tag{2.32}$$

and is an example of the rare helium fusion process. Then, lithium may appear by

$$^{7}Be + \beta^{-} \rightarrow {^{7}Li} \,, \tag{2.33}$$

and so on to higher nuclides.

An examination of the light-nuclide part of the Chart of the Nuclides suggests the possibility of additional collisional events. The suggestion of a nuclear "building-block" universe is thus evident.

Closed reaction cycles have also been proposed of which the Carbon Cycle is particularly well known:

$$
\begin{aligned}
{}^{12}C + {}^{1}H &\rightarrow {}^{13}N + 1.9\ \text{MeV} \\
{}^{13}N &\rightarrow {}^{13}C + \beta^{+} + \nu + 1.5\ \text{MeV} \\
{}^{13}C + {}^{1}H &\rightarrow {}^{14}N + 7.6\ \text{MeV} \\
{}^{14}N + {}^{1}H &\rightarrow {}^{15}O + 7.3\ \text{MeV} \\
{}^{15}O &\rightarrow {}^{15}N + \beta^{+} + \nu + 1.8\ \text{MeV} \\
{}^{15}N + {}^{1}H &\rightarrow {}^{12}C + \alpha + 5.0\ \text{MeV}\ .
\end{aligned}
\tag{2.34}
$$

This closed sequence of reactions may be collectively represented by

$$4\,{}^{1}H \rightarrow {}^{4}He + 2\,\beta^{+} + 2\nu \ , \tag{2.35}$$

and the associated mass decrement for the entire cycle, Eq. (2.34), is therefore

$$\Delta m = [\, m_{r,\alpha} + 2\, m_{r,\beta} - 4\, m_{r,p} \,]\ . \tag{2.36}$$

Thus, while heavier elements are formed, energy is released. Note that, herein, the neutrino, like the gamma ray photon, possesses zero rest mass.

The above reactions and reaction sequences represent but a very small part of what is called nucleosynthesis and nucleogenesis. In this science, one seeks to interpret known abundances of various nuclides and, together with measured nuclear reaction rate parameters and postulated temperature-pressure conditions in stellar media, reconstruct conceivable histories of the evolution of matter in the universe. Current views are that, except for some secondary effects, sequences of nuclear reactions beginning initially with pure hydrogen are the prime causes of all elements and that the associated mass decrement multiplied by c^2 is the dominant source of stellar energy. The matter-energy dynamic operative at the small scale local level also functions on the grand scale of the universe and its lifetime.

Problems

2.1 Consider the similarities between the following:

a) various atoms combine to form different molecules and

b) various nucleons combine to form different nuclides.

From the perspective of combinatorics, consider the number of distinct kinds of molecules and the number of distinct nuclides which are, in principle, conceivable.

2.2 What are the energies of gamma rays emitted by an excited ^{10}B nucleus?

2.3 Complete the following reaction by identifying the missing particle:

$^{3}H + ^{2}H \rightarrow [\ \ ] + ^{4}He$

$^{14}C \rightarrow ^{14}N + [\ \ ]$

$^{6}Li + n \rightarrow ^{3}H + [\ \ ]$

$^{137}I \rightarrow [\ \ ] + \beta^{-}$

2.4 Use the Chart of the Nuclides and conceive of feasible nuclear processes leading to the "manufacturing" of gold.

2.5 Compute the Q-value for Eq. (2.35) and compare your value to that from a sum of the Q-values of Eq. (2.34). Should there be a difference?

CHAPTER III

NUCLEAR DECAY

The spontaneous rearrangement of a nuclear configuration represents a discrete transition. For a large collection of radioactive nuclei, the collective change with time may be well described by methods of dynamical analysis. We examine some of these time dependent features and establish both analytical formulations and graphical depictions.

3.1 Radioactive Decay

The discovery at the turn of the century that certain naturally occurring substances emitted invisible and penetrating radiation, proved to be of pivotal importance to the development of nuclear science. In such events, the nucleus spontaneously undergoes a nucleon rearrangement which may include ejection of some nuclear particles and/or gamma ray photons; concurrently, a mass decrement occurs with its associated energy appearing in the motion of the reaction products.

In order to introduce sufficient generality, we consider the following description. The simplest radioactive decay process consists of a nucleus in its ground state, to be represented by "a", which spontaneously undergoes a rearrangement of its nucleons and ejects some particles with the residual nucleus "b" in its ground state. We write this transition as

$$a \rightarrow b + d\,, \tag{3.1}$$

where "d" is the emitted particle. It may also occur that the transition results in an excited state of the nucleus "b" which then de-excites, in a very short time, to its ground state. These two decay modes are suggested in Fig. 3.1 below.

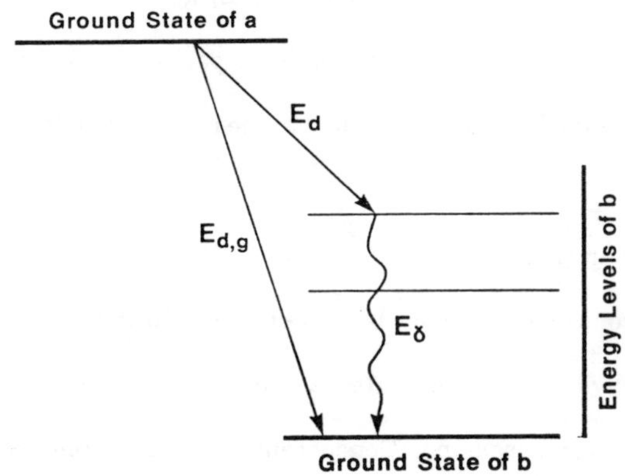

Fig. 3.1: Illustration suggesting two possible ways of how a radioactive nucleus "a" may spontaneously decay to another nucleus "b".

The energetics of the process are described by mass-energy conservation as discussed in the previous chapter. The mass decrement between the initial and final states is given by

$$\Delta m = (m_b + m_d) - m_a , \tag{3.2}$$

where m_a, m_b and m_d are the rest masses of the particles "a", "b" and "d". The energy thus released is the Q-value of the reaction

$$Q_{ab} = -(\Delta m)\, c^2 . \tag{3.3}$$

and is shared according to the laws of momentum conservation by all participating particles and electromagnetic radiation. In terms of the energy assignment of Fig. 3.1, we extend the analysis of matter-energy conservation of Chapt. I, the definition of the Q-value, Eq. (2.28), and the discussion which led to Fig. 2.4 to write

$$\begin{aligned} Q_{ab} &= E_{d,g} + E_{recoil} \\ &= E_d + E_\gamma + E_{recoil} \,. \end{aligned} \tag{3.4}$$

Here E_d and E_γ are the energies of the emitted particle and the gamma ray photon respectively; an amount of energy is assigned to the recoil of particle "b".

3.2 Particle Accounting

Important consequences may be associated with a large number of radioactive nuclear species. Evidently, the material composition changes, with one species diminishing and another increasing. Concurrently, energy is released. A time-dependent accounting of both matter and energy is thus essential. Appendix B provides some useful background information on nuclear particle accounting and should be studied.

Consider therefore an initial number of $N_{a,o}$ radioactive nuclei of type "a" in a unit volume at an arbitrary reference time $t = 0$. Our interest is to determine an adequate dynamical description applicable for all subsequent times. At the microscopic level, we have for a typical event, a mass-energy description given by

$$a \rightarrow b + Q_{ab} \tag{3.5}$$

and we want to know the number density of a-type nuclei at any time, to be represented by $N_a(t)$, the number density of b-type nuclei at any time, $N_b(t)$, as well as the power density released at any time, $P_{ab}(t)$; the emitted gamma ray is of little interest for now.

Discussion / Analysis 3.1

Alpha emitters with $Q \sim 5\,\text{MeV}$ are frequently used as power sources in space satellites.

e.g. $^{210}Po \longrightarrow {}^{206}Pb + \alpha, \quad \tau_{1/2} = 138\,d$

Estimate the radioisotope disintegration rate at time t when 1 kW of power is required.

Contains N_a number of radioisotopes.

$$P_\alpha = R_{-a}\, Q_\alpha$$

Q_α — MeV/reaction, 5 MeV this case

R_{-a} — disintegration rate of N_a, s^{-1}

P_α — thermal power, MeV/s ⟺ J/s, Table A.3

$$\therefore R_{-a} = \frac{P_\alpha}{Q_\alpha} = \frac{10^3\ \text{J/s}}{5\ \text{MeV} \times (1.6\times10^{-19}\ \text{J/eV})} = \boxed{1.25\times10^{15}\ s^{-1}}$$

Power decreases with time

$$P_\alpha(t) = R_{-\alpha}(t)\, Q_\alpha = \lambda_\alpha N_a(t)\, Q_\alpha = \lambda_a \left(N_{a,0}\, e^{-\lambda_a t} \right) Q_\alpha$$

To think about:

Why are alpha emitters particularly suitable for radioisotope power applications?

Equation (3.5) is a short-hand notation for a nuclear matter-energy transformation. The dynamics of the process at our level of interest are -- in view of Appendix B -- given by the associated dynamical equations for the densities $N_a(t)$ and $N_b(t)$ as well as for the power density $P_{ab}(t)$:

$$\frac{dN_a}{dt} = R_{+a} - R_{-a}, \tag{3.6a}$$

$$\frac{dN_b}{dt} = R_{+b} - R_{-b}, \tag{3.6b}$$

$$P_{ab}(t) = R_{-a} Q_{ab} = R_{+b} Q_{ab}. \tag{3.6c}$$

Here, again $R_{\pm i}$ refers to the reaction density rates which lead to a gain/loss of the i'th nuclear specie; note also the two equivalent expressions for the associated nuclear decay power density.

Further, for our case of interest, Eq. (3.5), we have

$$R_{+a} = R_{-b} = 0, \tag{3.7a}$$

$$R_{-a} = R_{+b} \neq 0, \tag{3.7b}$$

so that only one of R_{-a} or R_{+b} needs to be determined. We choose to determine R_{-a}.

Conceptual considerations demand that

$$R_{-a} = \left(-\frac{dN_a}{dt}\right)_{decay}, \tag{3.8}$$

where $(-dN_a/dt)_{decay}$ identifies specifically the radioactive decay process. Of interest then is the determination of a differential expression which adequately accounts for the uniqueness of the radioactive decay process. Consider the following intuitive approach.

Observations have shown that the fractional change of a statistically large collection of radioactive nuclei, $\Delta N_a/N_a$, per unit time interval Δt can be characterized by a constant for a given type of radioactive nuclide. That is, for the a-type nuclides, we write

$$\left(\frac{-\Delta N_a/N_a}{\Delta t}\right)_{decay} \sim \text{constant} = \lambda_a . \tag{3.9}$$

Here, λ_a is that particular decay constant and chosen to be positive by the inclusion of the negative sign (since ΔN_a is evidently negative and Δt is positive). Rearrangement of Eq. (3.9) yields

$$\begin{aligned}\left(-\frac{\Delta N_a}{\Delta t}\right)_{decay} &= \lambda_a N_a(t) \\ &= R_{-a} .\end{aligned} \tag{3.10}$$

Our important dynamical equation for $N_a(t)$, Eq. (3.6a), is therefore

$$\frac{dN_a}{dt} = 0 - \lambda_a N_a(t), \qquad N_a(0) = N_{a,0}, \tag{3.11}$$

with the initial conditions also indicated. Its solution is evidently

$$N_a(t) = N_{a,o} \exp(-\lambda_a t) . \tag{3.12}$$

Thus, the number of radioactive nuclei decreases exponentially with time.

We suggest in the problem set at the end of this chapter, another approach which leads to Eqs. (3.11) and (3.12).

According to Eq. (3.7b), we write the dynamical equation for $N_b(t)$,

$$\begin{aligned}\frac{dN_b}{dt} &= \lambda_a N_a(t) \\ &= \lambda_a N_{a,0} \exp(-\lambda_a t), \qquad N_b(0) = 0 .\end{aligned} \tag{3.13}$$

With the initial condition also indicated, we obtain by integration

$$N_b(t) = N_{a,0} \{1 - \exp(-\lambda_a t)\} . \tag{3.14}$$

We depict both $N_a(t)$ and $N_b(t)$ in the upper part of Fig. 3.2.

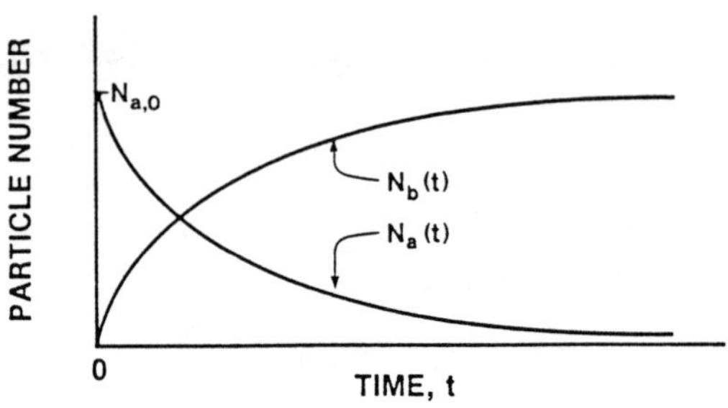

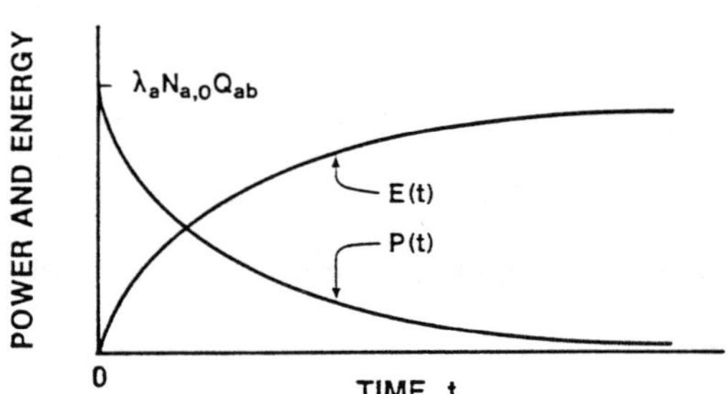

Fig. 3.2: The upper part shows the time variation of the a-type and b-type particle densities associated with radioactive decay; the lower part shows the associated power production and cumulative energy release.

The power released at any time of the decay process, Eq. (3.6c), is therefore

$$\begin{aligned} P_{ab}(t) &= R_{-a} Q_{ab} \\ &= \lambda_a N_a(t) Q_{ab} \\ &= \lambda_a Q_{ab} N_{a,0} \exp(-\lambda_a t) . \end{aligned} \tag{3.15}$$

which also decreases exponentially, Fig. 3.2.

Finally, the total cumulative energy thus released up to an arbitrary time follows from the definition

$$P_{ab}(t) = \frac{dE_{ab}}{dt}, \tag{3.16}$$

and integration

$$\int_{E_{ab}(0)=0}^{E_{ab}(t)} dE_{ab} = \int_{t=0}^{t} P_{ab}(t)\, dt. \tag{3.17}$$

Hence

$$\begin{aligned} E_{ab}(t) &= \int_0^t \lambda_a Q_{ab} N_{a.0} \exp(-\lambda_a t)\, dt \\ &= \lambda_a Q_{ab} N_{a.0} \int_0^t \exp(-\lambda_a t)\, dt \\ &= N_{a.0} Q_{ab} \{1 - \exp(-\lambda_a t)\}. \end{aligned} \tag{3.18}$$

We depict both P(t) and E(t) in the lower part of Fig. 3.2.

3.3 Decay Characterizations

The exponential decrease in the initial number of radioactive nuclei provides for some convenient characterizations. For example, the time required until only $\frac{1}{2}$ of the initial $N_{a,o}$ radioactive nuclei remain -- to be represented by $\tau_{1/2}$ and called half-life -- is defined by

$$N_a(\tau_{1/2}) = \tfrac{1}{2} N_{a,o}, \tag{3.19}$$

and determined with the aid of Eq. (3.12) for $t = \tau_{1/2}$:

$$N_a(\tau_{1/2}) = N_{a,o} \exp(-\lambda_a \tau_{1/2}). \tag{3.20}$$

Equating both these expressions, Eq. (3.19) and Eq. (3.20), and solving for $\tau_{1/2}$ gives

$$\tau_{1/2} = \frac{\ell n\,(2)}{\lambda_a} \simeq \frac{0.693...}{\lambda_a}, \tag{3.21}$$

and is independent of any reference density of particles. That is, the time required for a 50% decrease in the number of radioactive nuclei is always $\sim 0.693/\lambda_a$.

Further, the exponential decrease in the number of radioactive nuclei with time implies that -- like "living" things -- some of the nuclei will live for a short time and others for a long time. A mean-life, τ, would therefore be of some intuitive use in the characterization of radioactive species. By definition, the mean-life is found from

$$\bar{\tau} = \frac{\int_0^\infty t\, p(t)\, dt}{\int_0^\infty p(t)\, dt} , \qquad (3.22)$$

where p(t) is the probability density of decay of a nuclide at time t given it was present at t = 0. We construct this distribution function by the following argument. We take

$$\begin{aligned} p(t)\, dt &= \begin{pmatrix} \text{Probability of} \\ \text{survival up to t} \end{pmatrix} \begin{pmatrix} \text{Probability of decay} \\ \text{within t + dt} \end{pmatrix} \\ &= \left[\frac{N_{a,o} \exp(-\lambda_a t)}{N_{a,o}} \right] \left[- \frac{d N_a}{N_a} \right] \\ &= \left[\exp(-\lambda_a t) \right] \left[\lambda_a dt \right] , \end{aligned} \qquad (3.23)$$

and hence,

$$p(t) = \lambda_a \exp(-\lambda_a t) . \qquad (3.24)$$

The mean-life for radioactive nuclei of type "a" is therefore

$$\begin{aligned} \bar{\tau}_a &= \frac{\int_0^\infty t\, \lambda_a \exp(-\lambda_a t)\, dt}{\int_0^\infty \lambda_a \exp[-\lambda_a t]\, dt} \\ &= \frac{1}{\lambda_a} . \end{aligned} \qquad (3.25)$$

It has become customary to use the expression "activity" for the radioactive disintegration rate without regard to atomic densities. This activity is therefore given by

$$\text{Activity} = R^0_{-a} = \lambda_a N^0_{a,0} \exp(-\lambda_a t), \tag{3.26}$$

with the superscript in $R^0{}_{-a}$ and $N^0{}_{a,o}$ now representing volume independent quantities; that is, $R^0{}_{-a}$ has units of s^{-1}, and $N^0{}_{a,o}$ is a dimensionless particle number. Hence, activity has units of reciprocal time. The specific name in the International System of units for activity is the becquerel, represented by the symbol Bq:

$$1 \text{ becquerel (Bq)} = 1 \text{ disintegration per second} . \tag{3.27}$$

An older unit, still in use, is the curie (Ci). It is related to the becquerel by

$$1 \text{ Ci} = 3.7 \times 10^{10} \text{ Bq} = 37 \text{ GBq} . \tag{3.28}$$

3.4 Nuclide Densities

Evaluation of radioactive decay rates, as well as nuclear reaction interaction rates, requires the specifications of nuclear densities, N. The determination of this quantity is possible with the aid of some definitions and properties of matter as follows:

(1) The masses of all atoms are referenced to the ^{12}C atom such that an atomic mass unit (u) is defined by

$$1 \text{ u} = (1/12) \text{ mass of } ^{12}\text{C atom} ; \tag{3.29a}$$

e.g. 1 ^{12}C atom → 12 u; 1 aluminum atom → 26.98 u; average of natural uranium atoms → 238.03 u.

(2) The amount of matter is given by a quantity called the mole and defined as

1 mole = amount of a substance whose mass in grams is numerically equal to its atomic mass ; (3.29b)

e.g. 1 mole ^{12}C → 12 g; 1 mole aluminum → 26.98 g; 1 mole of natural uranium → 238.03 g.

Discussion / Analysis 3.2

Consider a ^{60}Co medical therapy facility. When the radioisotope activity has decreased to 40% of its initial activity, the exposure times become too long and the source needs replacing. How long is this period?

shield enclosure

source : $N_c(t)$, $\tau_{1/2} = 5.24$ yrs (Chart of Nuclides)

Table for patient

Source activity :

Initial $t = 0$ $\quad R_{-c}(0) = \lambda_c N_c(0)$

Replacement $t = t_1$ $\quad R_{-c}(t_1) = \lambda_c N_c(t_1)$

Replacement criterion :

$$R_{-c}(t_1) = 0.4\, R_{-c}(0)$$

i.e. $$\lambda_c N_c(t_1) = 0.4\, \lambda_c N_c(0)$$

$$\lambda_c N_c(0) e^{-\lambda_c t_1} = 0.4\, \lambda_c N_c(0)$$

Solve for t_1 : $t_1 = \frac{1}{\lambda_c} \ln\left(\frac{1}{.4}\right) = \frac{5.24}{\ln(2)} \ln\left(\frac{1}{.4}\right) = \boxed{6.9 \text{ yrs}}$

To think about:

How would one determine the increasing exposure time, with increasing age of radioisotope, for same therapy effect?

(3) The number of atoms in 1 mole of a substance is a constant, N_A, called the Avogadro constant; experimentally, it has been found to be

$$N_A = 6.022 \times 10^{23} \text{ atoms/mole} . \qquad (3.29c)$$

From the above, we conclude that any matter of bulk density $\rho(\text{g/cm}^3)$ and containing a substance characterized by M_u (g/mole) will evidently contain ρ/M_u (moles/cm^3). With each mole containing N_A atoms (= nuclides) we obtain a nuclear density $N(\text{cm}^{-3})$ of

$$N = \frac{\rho N_A}{M_u} \left(\frac{\text{nuclides}}{\text{cm}^3} \right) . \qquad (3.30)$$

To within an approximation suggested above, $M_u \simeq A$ nucleons in a nuclide. Typical nuclide densities of solids are therefore in the 10^{22} nuclides/cm^3 range, while gases are less dense by $\sim 10^3$, Appendix A, Table A.5.

3.5 Radioisotope Production

Radioisotopes can be selectively produced in nuclear reactors or in nuclear accelerators. A suitable parent material composed of, say, "a"-type atoms is bombarded by either neutrons or ions, to be designated by "b", and transmuted according to the collisional process

$$a + b \rightarrow d \rightarrow e + \ldots . \qquad (3.31)$$

Here, "d" is the desired radioisotope and "e" is its daughter decay product. After a sufficient number of these radioisotopes "d" are produced, the sample is then withdrawn.

The time-dependent analysis of N_d must include considerations of its production rate and decay rates based on the particle accounting. This involves the dynamical equation

$$\frac{d N_d}{dt} = R_{+d} - R_{-d} , \qquad (3.32)$$

to be specified for the given time interval of interest. Thus, while the sample is in the nuclear device, we take

$$R_{+d} = R_{ab} , \tag{3.33}$$

as a constant. Then, the process which concurrently contributes to the removal of "d" is its own decay rate

$$R_{-d} = \lambda_d N_d . \tag{3.34}$$

If these are the principal processes which affect the density of N_d, we obtain for N_d the first-order differential equation

$$\frac{d N_d}{dt} = R_{ab} - \lambda_d N_d , \qquad N_d(0) = 0 , \tag{3.35}$$

with R_{ab} and λ_d as constants. The solution of this differential equation is

$$N_d(t) = \frac{R_{ab}}{\lambda_d} [1 - \exp(-\lambda_d t)], \qquad 0 \le t \le t_1 , \tag{3.36}$$

where t_1 is the irradiation time.

At $t = t_1$, the sample is withdrawn and the radioisotope concentration is affected only by its own decay

$$\frac{d N_d}{dt} = - \lambda_d N_d , \qquad t > t_1 , \; N_d(t_1) = N_{d,1} , \tag{3.37}$$

valid only for $t > t_1$; the initial condition $N_{d,1}$ is based on Eq. (3.36):

$$N_d(t_1) = N_{d,1} = \frac{R_{ab}}{\lambda_d} [1 - \exp(-\lambda_d t_1)] . \tag{3.38}$$

Solving Eq. (3.37) by inspection followed by substitution using Eq. (3.38), we obtain

$$\begin{aligned} N_d(t) &= N_d(t_1) \exp[-\lambda_c(t - t_1)] \\ &= \frac{R_{ab}}{\lambda_d} \{1 - \exp(-\lambda_d t_1)\} \exp[-\lambda_d(t - t_1)] , \qquad t \ge t_1 . \end{aligned} \tag{3.39}$$

With Eq. (3.36) valid for $0 \le t \le t_1$ and Eq. (3.39) for $t \ge t_1$, we have arrived at the complete description of the time variations of the isotope "d"; a graphical depiction is shown in Fig. 3.3.

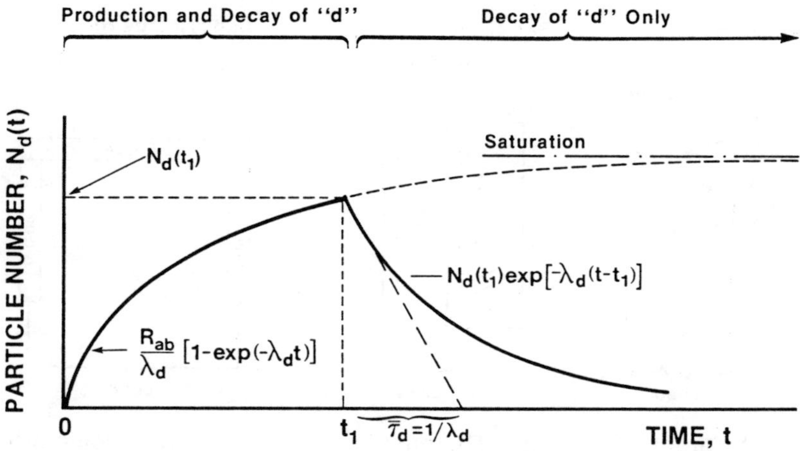

Fig. 3.3: Time variation of the number of isotopes N_d. From $t = 0$ to $t = t_1$, both production and removal of "d" occur; after $t = t_1$, only removal by radioactive decay occurs.

3.6 Isotope Power

The recognition that energy is released and power is generated when a nucleus decays, suggests that methods and devices be developed which could extract this energy in a particularly useful form such as electricity. Three distinct devices have been conceived and tested for such a nuclear-to-electrical conversion of energy.

A particularly simple radioisotope power device consists of an alpha or beta emitter located in the center of an evacuated spherical chamber. The charged particles are collected by the outer shell with the result that a potential difference is established to sustain a load. Figure 3.4a depicts such a direct charging nuclear battery.

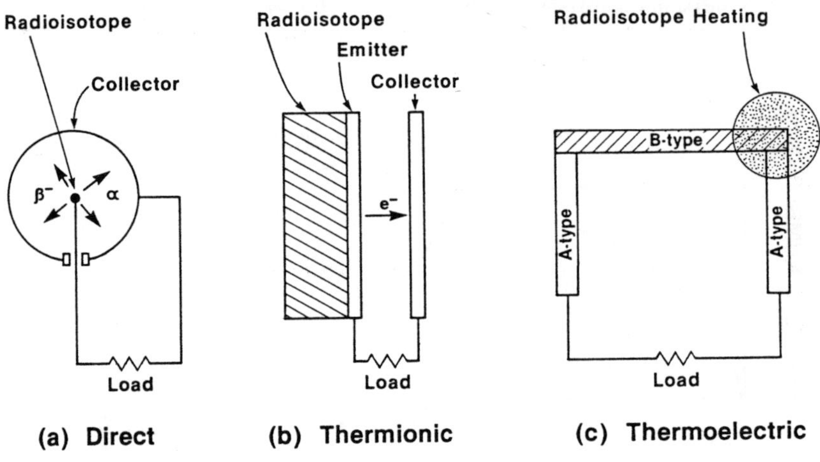

Fig. 3.4: Graphical characterization of three types of isotope power sources.

A thermionic power source is based on the principle that the kinetic energy of the decay particles and associated decay energy heats up the material in which the isotope is contained. This thermal heat is then transferred by conduction to a suitable material which possesses the property of "boiling-off" electrons at elevated temperatures. Again, a current is established to which a load can be attached, Fig. 3.4b.

Finally, a third type of device is called a thermoelectric radioisotope power source and is based on the Seebeck effect; this effect describes the voltage difference in a circuit of dissimilar metals if the junctions are at different temperatures. A radioisotope medium which is particularly effective in generating thermal heat from its decay process is attached to one of the junctions, with the other junction held at the lowest possible temperature, Fig. 3.4c.

The thermionic and thermoelectric power devices have been found useful in specialized applications such as implantable heart pacers, lunar-based power sources, and power sources for isolated and automated weather stations.

Problems

3.1 The activity of a short lived isotope ($\tau_{1/2}$ = 28 s) has been measured to be 10^{11} Bq. How long will it take for the activity to attain 10^5 Bq?

3.2 The present relative abundance of natural uranium is 0.9928 of ^{238}U ($\tau_{1/2} = 4.47 \times 10^9$ y) and 0.0072 of ^{235}U ($\tau_{1/2} = 7.02 \times 10^8$ y). What were the abundances 10^6 years ago?

3.3 During an expected two-year lunar-based operating time, the power output from the radioisotope source must not decrease by more than 15%. What restrictions does this impose on the lower limits of half-life on the chosen isotope?

3.4 Consider the radioisotope chain a $\rightarrow$ b $\rightarrow$ d , with initial conditions $N_a = N_{a,0}$, $N_b(0) = 0$ and $N_d(0) = 0$. The product d is stable. Obtain expressions for $N_a(t)$, $N_b(t)$ and $N_d(t)$. Sketch the time dependent activities.

3.5 How many disintegrations per second are associated with (a) a gram of tritium? (b) a gram of radium?

3.6 The radioactive decay rate expression, Eqs. (3.11) and (3.12), can also be derived from the premise that the conditional probability of nuclear disintegration in Δt after t is given by $P = \lambda \Delta t$ where λ is the decay constant. Use a physical interpretation for P and known differential/integral relationships between a probability and a probability density.

CHAPTER IV

RADIATION ASSESSMENT

Radiation is part of our environment somewhat like air and gravity. We consider first some phenomenological aspects of radiation and subsequently formulate selected relations for its description.

4.1 Radiation Environment

Radiation is a phenomenon characterized by a capacity to penetrate matter and interact with the constituent molecules, atoms and nuclei. The entire electromagnetic spectrum -- from very short gamma-rays through the visible light component to very long radio waves -- is part of our radiation environment.

Radiation in our environment may be usefully classified according to its source of origin. Three such classifications may be conveniently identified:

1. natural cosmic radiation,
2. natural terrestrial radiation,
3. man-made radiation.

Cosmic radiation consists of protons, pions, muons and other sub-nuclear species as well as stellar and interstellar photons. The intensity of this radiation increases noticeably with elevation.

Naturally occurring terrestrial radiation results from the decay of long-lived radioisotopes in the earth. Some of these invariably find their way into the food chain, into building materials, and other man-made products. Additionally, radioisotopes exist in the air as a consequence of the interstellar radiation bombardment and also mix with food products to enter the human body.

Man-made radiation is associated with various devices such as medical X-ray facilities, gamma-ray therapy installations, accelerators, nuclear reactors, television, household smoke detectors, and numerous other objects.

4.2 Descriptive Concepts

Nuclear radiation may be graphically depicted by either a "particle-like" or "wave-like" representation. The demonstration that a given radiation phenomenon may exhibit both a wave-property and a corpuscular property in a given experiment, provides the assurance that either model is valid and a choice of one over another is primarily a matter of convenience and context.

If we think of radiation as particles -- of which protons and neutrons are suitable examples -- then we may assign a rest mass m_r and a speed v to this corpuscular phenomenon, Fig. 4.1a. The total energy of such a particle is fully defined in terms of its rest mass energy and its kinetic energy:

$$E = m_r c^2 + E_k . \tag{4.1}$$

Choosing a wave description, which is particularly useful for a gamma-ray wave packet, leads to a description as suggested in Fig. 4.1b. Its total energy is given by the relation

$$E = h\nu , \tag{4.2}$$

where h is Planck's constant (Appendix A) and ν is the photon frequency.

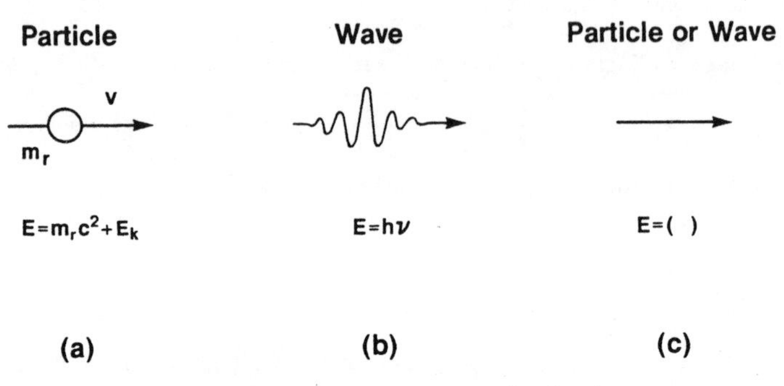

Fig. 4.1: Graphical depictions of radiation.

Regardless of whether one imagines these entities as waves or particles -- indeed, one might be tempted to become inventive and introduce the generic label of "waveticle" -- the meaning of density of these entities is self-evident: it simply represents the statistically expected number in a unit volume and represented by the scalar $N_{()}$ possessing units of cm^{-3}, Fig. 4.2a. However, since each entity possesses some speed and some direction of motion, care needs to be exercised at the next level of descriptive interest because directional considerations now become important. Consider, then, the following cases in which, for reasons of literal simplicity, we will be referring specifically to the motion of particles.

Supposing there exists a population of highly mobile a-type particles of density N_a moving about in relatively more dense but less mobile background particles. Assume next that all of the a-type particles possess a nearly constant speed v_a but that all directions of motion are equally likely; that is, the a-type particles represent an essentially monoenergetic and isotropic population. The product $N_a v_a$ then represents a measure of particle motion in the unit volume. The symbol ϕ_a is used for this product

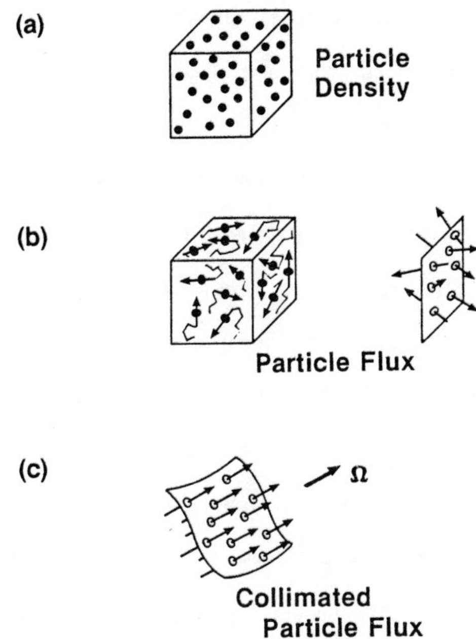

Fig. 4.2: Depictions of a particle density (a), two illustrations of a (isotropic) particle flux (b), and a collimated (directional) particle flux (c).

$$\phi_a = N_a \, v_a \,, \tag{4.3}$$

and generally called the particle flux; we may qualify this quantity for the situation described by calling it the isotropic particle flux.

The units of ϕ_a provide for two intuitive and frequently helpful interpretations. With N_a in units of cm^{-3} and v_a in units ($cm \; s^{-1}$), ϕ_a thus possesses units of (cm) ($cm^{-3} \; s^{-1}$) and may be viewed as representing the total path length of the a-type particle in the unit volume during one unit of time. Further, the units also reduce to ($cm^{-2} \; s^{-1}$), and ϕ_a may also be interpreted as the total number of penetrations through some unit area in a unit of time by the a-type particles. The former interpretation is expected to be useful when analysing density phenomena, such as reaction density rates, while the latter may be useful for surface effects, such as radiation leakage. We suggest a graphical depiction of these two interpretations in Fig. 4.2b.

Suppose, however, that the a-type particles possess a non-isotropic directional distribution! Taking an extreme case, suppose they are all moving in the same direction and thus constitute a collimated beam; or, what occurs more frequently, there exists some non-isotropic directional distribution but our interest is only in those particles moving in a direction within the differential solid angle $d\Omega$ about $\mathbf{\Omega}$. We may then write

$$\phi_{a,\Omega} = (N_a \, v_a)_\Omega \,, \tag{4.4}$$

where the subscript Ω is to signify that N_a and v_a inside the bracket applies only to those particles moving within $d\Omega$ in the $\mathbf{\Omega}$ direction. Numerically and physically, $\phi_{a,\Omega}$ now identifies only the number of particles per unit area which are moving in the direction Ω, Fig. 4.2c. Directional, or collimated flux, are terms frequently used to distinguish $\phi_{a,\Omega}$ from the isotropic flux introduced in Eq. (4.3). It is therefore important to note the context in which the term flux and the symbol ϕ is used.

The last concept we introduce here is that of the radiation energy retained in an object exposed to radiation. That is, as radiation passes into a medium, the photons or particles interact in a variety of ways with the molecules, atoms, and nuclides and in so doing deposit some or all their total energy. The energy thus deposited per unit mass is called the radiation energy absorbed dose and frequently represented by the symbol D. This phenomenon is generally quantified in units of Gy (abbreviation for gray). In the SI system, 1 Gy = 1 J/kg and is a scalar; the previous unit of dose was the rad (acronym for radiation absorbed dose) now defined by 1 rad = 0.01 Gy.

4.3 Geometric Effects

Of interest here is the effect of geometry on the collimated radiation flux. Material effects, that is, radiation interactions with matter, is for now assumed to be negligible; for most cases, air at standard temperature and pressure is adequate to simulate such circumstances.

For the case of a collimated radiation beam in free space, Fig. 4.3a, geometrical effects are evidently irrelevant. The radiation flux at x_1 in the $\mathbf{\Omega}$ direction, $\phi_\Omega(x_1)$, is identical to that at x_2

$$\phi_\Omega(x_1) = \phi_\Omega(x_2) \ , \tag{4.5}$$

because there exists no material in the interval to affect the radiation beam.

A different situation arises if the radiation emerges isotropically from a sufficiently small region to be considered a point source, Fig. 4.3b. If the source of radiation emits S photons or particles per second then, in the absence of any radiation-matter interaction effects, all of these radiations must penetrate through the surface of a sphere of radius r. Taking $\phi(r)$ as the flux of radiations per unit area per unit time -- diverging outward from a point in this case -- and the area of the sphere at distance r is $4\pi r^2$, requires therefore

$$S = \phi(r)\, 4\pi r^2 \, . \tag{4.6}$$

Hence, for an isotropically emitting radiation point source, the radiation flux at a distance r is

$$\phi(r) = \left(\frac{1}{4\pi}\right)\frac{S}{r^2} \, . \tag{4.7}$$

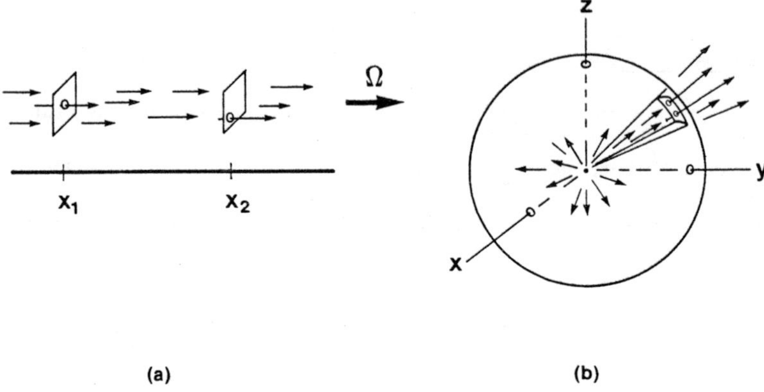

Fig. 4.3: Illustration of a collimated radiation beam (a), and a divergent beam from a point radiation source (b).

This result illustrates an important aspect of geometric shielding from a point source of radiation: a distance twice as far from the source reduces the radiation flux four-fold.

4.4 Material Effects

The effect of matter on radiation is governed by the interactions which occur. Here it is necessary to distinguish between the various radiation-matter interactions. Of primary interest is the penetration of gamma-ray photons and of neutrons and we will therefore consider each separately.

When a gamma-ray photon enters a medium, any one of four types of events may occur:

(a) Photoelectric Effect: a gamma-ray photon deposits all of its energy in the process of ejecting an electron from the atomic shell of an atom.

(b) Compton Effect: a gamma-ray photon deposits part of its energy in the process of ejecting an electron from the atomic shell of an atom and then continues with reduced energy and different direction.

(c) Pair-Production: a gamma-ray photon transforms itself spontaneously into an electron and positron when near a nucleus.

(d) Transmission: a gamma-ray photon passes through the material object without collision.

We depict these processes in Fig. 4.4.

Several specific comments need to be made about the above processes. Note that the first two involve ionizing effects. Further, it is known that each process is very sensitive to the energy of the photon and material composition. Also, the transformation of penetrating radiation into particles such as electrons and positrons is of important shielding consequence because these secondary particles are more easily attenuated. Finally, it is the uncollided fourth component which is of particular computational interest in radiation shielding by material attenuation.

The interaction of neutrons in matter is governed by the important feature that neutrons do not possess a charge and hence may pass unimpeded through the electron cloud surrounding nuclei and interact directly with the nucleus. Among the important neutron-nucleus collisional effects are the following:

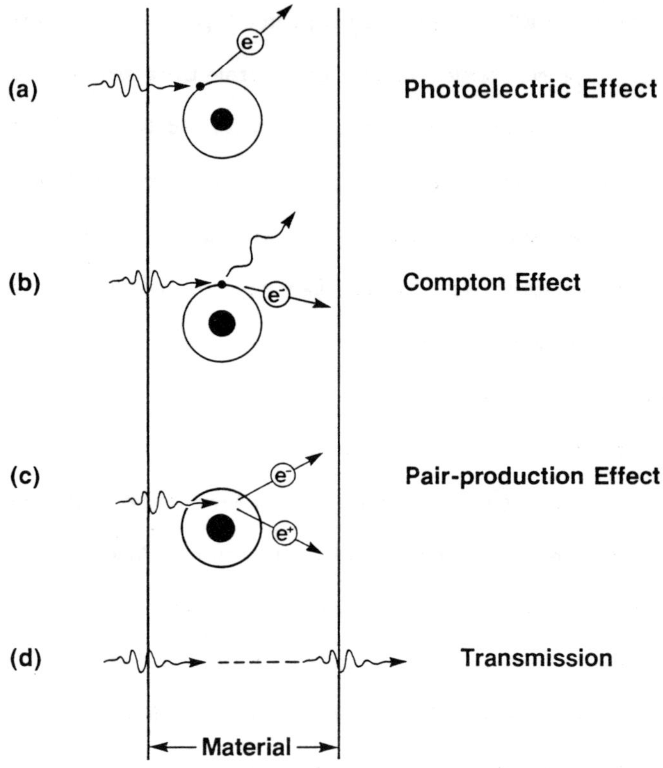

Fig. 4.4: Four possible events which may occur when a gamma ray photon enters a medium.

(a) Radiative Capture: a neutron is captured by a nucleus and gamma-ray(s) are emitted.

(b) Nucleon Displacement: a neutron enters a nucleus and one or more nucleons (neutrons, protons) or an alpha particle (^{4}He nucleus) are emitted.

(c) Neutron Scattering: a neutron scatters from a nucleus and continues with different direction and reduced kinetic energy.

(d) Neutron Transmission: a neutron may pass through a material object without collision of any type.

We depict these processes in Fig. 4.5. Compare this figure with Fig. 4.4 and note the distinctive collisional processes which, nevertheless, lead to similar consequences. Note also that both gamma rays and neutrons may pass through an object without collision.

The relative probability of the occurrence of any of the above neutron-nucleus processes is highly dependent upon the kinetic energy of the neutrons and the material. Unlike gamma-rays, however, the secondary particles resulting from collisional interactions together with multiple scattering effects renders the neutron attenuation frequently more difficult.

4.5 Material Attenuation

While the details of the interactions between gamma-rays and atoms, as well as between neutrons and nuclei, are indeed very different, it is remarkable that the fraction of uncollided radiation is given by an identical mathematical formulation though the magnitude of a parameter appearing therein may be significantly different. Hence we may use a common method of derivation for this transmitted radiation component.

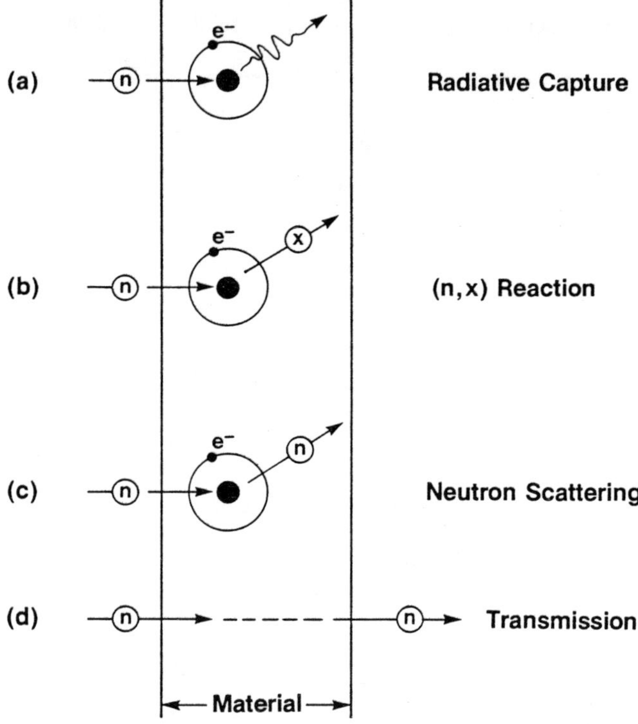

Fig. 4.5: Four possible events which may occur when a neutron enters a medium.

Consider a beam of collimated radiation of magnitude $\phi_{c,in}$, either neutrons or gammas, incident onto a homogeneous slab of material of thickness x_1; given $\phi_{c,in}$, we are interested in the exiting uncollided radiation beam $\phi_{c,out}$, Fig. 4.6.

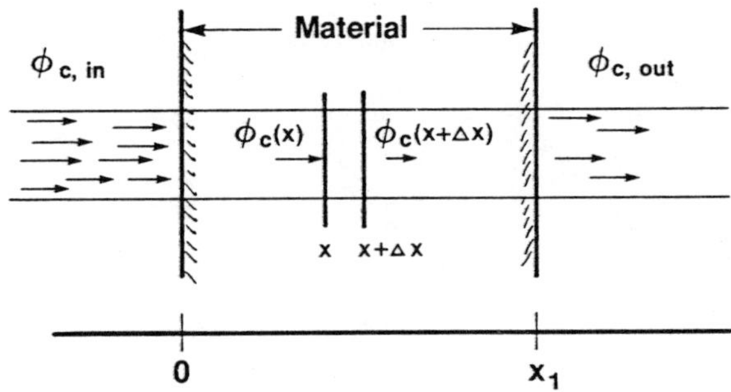

Fig. 4.6: Collimated radiation beam incident on a slab of material.

Evidently, $\phi_{c,in} > \phi_{c,out}$ with the difference consisting of those radiations, photons or particles, which have experienced any of the collisional effects suggested in Fig. 4.4 and Fig. 4.5; hence, $\phi_{c,out}$ represents only those of the initially collimated radiations which have not interacted at all. By analogy, it is a little like shooting a projectile at a grove of trees planted far apart; some of the bullets will ricochet about, some will be stuck in a tree trunk, but -- depending upon the distance separating the trees, the area covered by the grove, and the size of the projectile -- some projectiles will make it through without collision. It is the fraction of these uncollided projectiles we have an interest in because they determine, to a large extent, the "hazards" on the other side of the grove.

The slab of interest, Fig. 4.6, has been taken to be homogeneous and hence the material effects on the beam are similarly uniform everywhere. Consider then a very thin domain of thickness Δx for which $\phi_c(x)$ is the incident collimated uncollided part of the beam at coordinate x and $\phi_c(x+\Delta x)$ is the exiting uncollided beam at $x+\Delta x$. The fraction of the radiations in the beam which have collided -- and hence ceases to be part of the uncollided beam -- is evidently

$$\text{(collided fraction)} = -\frac{\Delta\phi_c}{\phi_c(x)} . \tag{4.8}$$

It is physically plausible to assume that for Δx sufficiently thin, say several interatomic distances, this fractional decrease is directly proportional to Δx:

$$-\frac{\Delta\phi_c}{\phi_c(x)} \propto \Delta x , \tag{4.9a}$$

or

$$-\frac{\Delta\phi_c}{\Delta x} \propto \phi_c(x) . \tag{4.9b}$$

Now, a proportionality relationship can always be converted into an equation by the introduction of a proportionality parameter appropriate for the phenomenon of interest. For our purposes, we choose to use the symbol μ, call it the radiation attenuation parameter, and write for Eq. (4.9b),

$$-\frac{\Delta\phi_c}{\Delta x} = \mu\,\phi_c(x) . \tag{4.10}$$

It is of importance to recognize that, while μ is a proportionality parameter, its magnitude is case specific. That is, for our case here, μ will have different numerical values depending upon the following:

(a) the form of radiation;

(b) the energy of the incident radiations;

(c) the composition of the slab.

Hence, we have a dependence suggested by the following:

$$\mu = \text{fc (type of radiation, energy of radiation, material of the object)} . \tag{4.11}$$

Continuing our derivation, we take Eq. (4.8) to the limit of an infinitesimally thin Δx and obtain

$$- \frac{d\phi_c}{dx} = \mu \, \phi_c(x) \, , \tag{4.12}$$

which is a first order differential equation valid everywhere in our material slab. Given a boundary condition, and here $\phi_{c,in} = \phi_c(0)$ is generally available, we finally write

$$\frac{d\phi_c}{dx} = - \mu \, \phi_c(x) \, , \qquad \phi_c(0) = \phi_{c,in} \, , \tag{4.13}$$

and solve by inspection

$$\phi_c(x) = \phi_{c,in} \exp(-\mu x) \, , \tag{4.14}$$

for x an arbitrary penetration distance. Finally then, with $\phi_c(x_1) = \phi_{c,out}$ we obtain the fraction of the beam passing through the slab of thickness x_1 without collision as

$$\frac{\phi_c(x_1)}{\phi_c(0)} = \frac{\phi_{c,out}}{\phi_{c,in}} = \exp(-\mu \, x_1) \, . \tag{4.15}$$

Table 4.1 lists attenuation parameters for selected materials at energies of common interest. The apparent increase in gamma-ray attenuation with increasing proton number, and hence also with increasing material density, is a general feature of both X-ray and γ-ray attenuation; such a trend, however, does not extend to neutron attenuation.

4.6 Radiation Shielding

The analysis above is well suited to the assessment of radiation shielding involving either collimated beams or point-sources of radiation.

Discussion / Analysis 4.1

Examine the application of thickness measurements using radioisotopes.

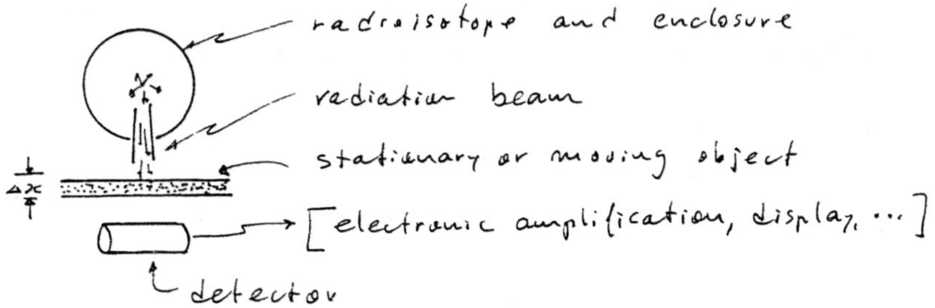

Radiation attenuation during counting interval

$$R_d(\Delta x) = R_d(0)\, e^{-\mu \Delta x}$$

- $R_d(\Delta x)$: (detector count rate with Δx in beam) $\propto$ (transmitted, uncollided radiation beam)
- $R_d(0)$: (detector count rate without slab)
- μ: attenuation parameter (case specific)
- Δx: slab thickness

Solving for Δx:

$$\Delta x = \frac{1}{\mu} \ln\left(\frac{R_d(0)}{R_d(\Delta x)}\right)$$

To think about:

Relate the above to the operation of a smoke detector based on use of radioisotopes.

Table 4.1

Listing of attenuation parameters for various materials

Material	Z	ρ (g/cm^3)	μ_γ (cm^{-1}) $E_\gamma = 1$ MeV	μ_n (cm^{-1}) $E_n = 0.025$ eV
Be: Beryllium	4	1.85	0.10	0.76
Mg: Magnesium	12	1.74	0.11	1.50
Al: Aluminum	13	2.70	0.17	0.10
Fe: Iron	20	7.87	0.47	1.15
Cu: Copper	29	8.96	0.51	0.99
Ti: Tungsten	74	19.2	1.23	1.16
Pb: Lead	82	11.3	0.77	0.36
U: Uranium	92	19.1	1.45	0.80

As a general example, consider the attenuation of a collimated beam of both gammas and neutrons. The unattenuated beam is composed of neutrons of current density $\phi_{c,n}$ (cm^{-2} s^{-1}) and gammas of current density $\phi_{c,\gamma}$ (cm^{-2} s^{-1}). Since these radiations are attenuated differently and independently by different materials, we consider a double-slab arrangement with $\mu_{n,1}$ and $\mu_{\gamma,1}$ as the attenuation parameters for the first slab and $\mu_{n,2}$ and $\mu_{\gamma,2}$ for the second, Fig. 4.7.

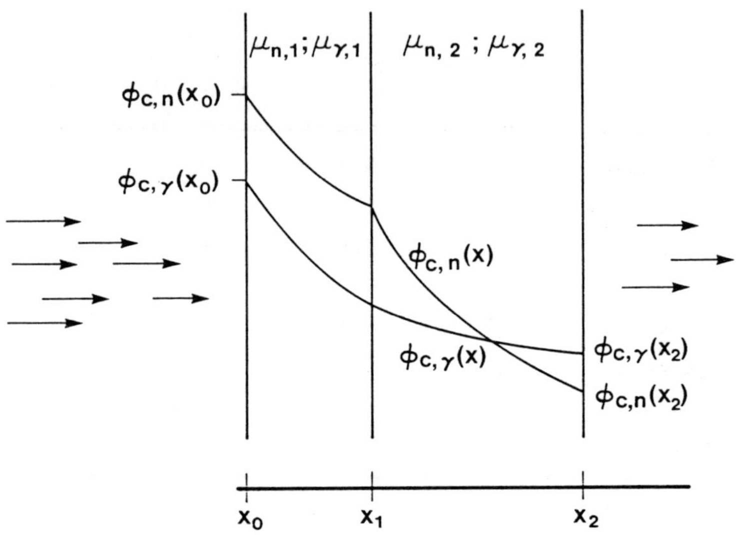

Fig. 4.7: Radiation attenuation of a beam in a two-media planar-geometry shield.

Our analysis now involves the repeated application of Eq. (4.14), taking care to apply it for the appropriate radiation and medium. For the first medium, we write

$$\phi_{c,n}(x_1) = \phi_{c,n}(x_0) \exp[-\mu_{n,1}(x_1-x_0)], \tag{4.16a}$$

$$\phi_{c,\gamma}(x_1) = \phi_{c,\gamma}(x_0) \exp[-\mu_{\gamma,1}(x_1-x_0)]. \tag{4.16b}$$

Furthermore, with the respective attenuation parameters known for the second medium, we write

$$\begin{aligned} \phi_{c,n}(x_2) &= \phi_{c,n}(x_1) \exp[-\mu_{n,2}(x_2-x_1)] \\ &= \phi_{c,n}(x_0) \exp[-\mu_{n,1}(x_1-x_0)] \exp[-\mu_{n,2}(x_2-x_1)], \end{aligned} \tag{4.17a}$$

$$\begin{aligned} \phi_{c,\gamma}(x_2) &= \phi_{c,\gamma}(x_1) \exp[-\mu_{\gamma,2}(x_2-x_1)] \\ &= \phi_{c,\gamma}(x_0) \exp[-\mu_{\gamma,1}(x_1-x_0)] \exp[-\mu_{\gamma,2}(x_2-x_1)]. \end{aligned} \tag{4.17b}$$

We can generalize this and write the attenuation for an N-layered shielding arrangement as

$$\frac{\phi_{c,out}}{\phi_{c,in}} = \prod_{i=1}^{N} \exp\left[-\mu_i (x_i - x_{i-1})\right]. \tag{4.18}$$

Shielding for a point radiation source involves both geometric and material effects. Consider a point source emitting S radiations per second which is surrounded by a spherical shell of inner radius r_1 and outer radius r_2, Fig. 4.8. At the inner surface we have a radiation flux of

$$\phi(x_1) = \frac{S}{4\pi r_1^2}, \tag{4.19}$$

as previously determined, Sec. 4.3. Then, if r_1 is sufficiently large and the shielded sphere sufficiently thin, we can consider this shield to have the effect of a planar attenuation slab in the radial direction. Hence, the radially directed radiation flux emerging from the shielded enclosure, $\phi(r_2)$, is

$$\phi(r_2) \simeq \phi(r_1) \exp\left[-\mu (r_2 - r_1)\right], \tag{4.20}$$

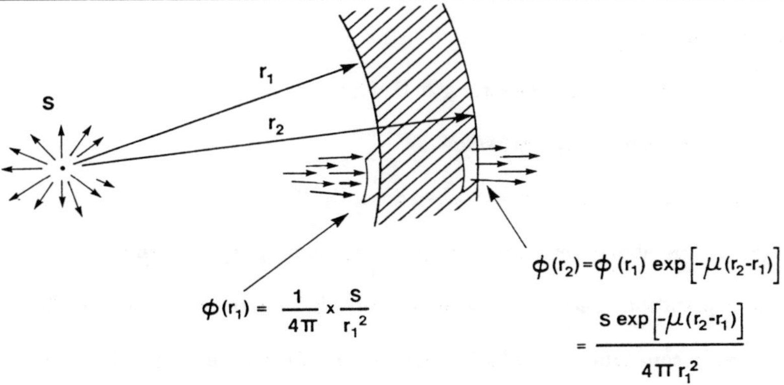

Fig. 4.8: Point radiation source surrounded by a shield.

where μ is the appropriate attenuation parameter. Substituting Eq. (4.17) herein yields specifically

$$\phi(r_2) \simeq \frac{S}{4\pi r_1^2} \exp[-\mu(r_2 - r_1)]. \tag{4.21}$$

Thus, an assessment of shielding requires a knowledge of the radiation source strength S, the kind of radiation and shielding material to specify the radiation attenuation parameter μ, and geometric parameters r_1 and r_2.

4.7 Radiation Exposure and Standards

Our preceding discussion has made it clear that radiation interaction in various materials leads to effects such as ionization, atomic displacement, and energy deposition. Both inorganic and organic materials may thereby change their properties and hence their functions. It is however, the radiation effect on human tissue and human organs which is of dominant importance. We consider here selected measures of radiation exposure and radiation standards.

Recall that we have identified several quantities useful in the characterization of radiation phenomena:

(a) radiation source strength, S (#/s)

(b) radiation flux, ϕ (#/cm^2-s)

(c) radiation energy absorbed dose, D(J/kg) or D(Gy).

While the above terms are adequate for some purposes, they are inadequate for biological applications because they do not contain any reference to the effectiveness in causing certain consequences in biological media. What is required is a multiplier which, if it multiplies the energy deposited in units of gray, Gy, yields an indication of biological consequences. This multiplier is represented by the symbol Q and called the "quality factor" and used to define a dose equivalent represented by the symbol Sv (for sievert)

$$1\ \text{Sv} = 1\ \text{Gy} \times \text{Q}\ . \tag{4.22}$$

With Q dimensionless, the units for Sv are the same as for Gy, i.e. J/kg. The older unit still in widespread use is the rem (roentgen equivalent man) and now defined as 100 rem = 1 Sv. The quality factor Q varies from a magnitude of 1 for gamma photons and beta particles to about 10-15 for alpha particles and neutrons of moderate to high energies.

With Gy and Q measurable quantities, it is possible to determine typical radiation doses to which humans are exposed. We itemize some normal and abnormal exposures in Table 4.2. Persons who frequently work in a radiation environment -- and this includes X-ray technicians, dentists, nuclear reactor operations staff, radiologists, industrial radiographers, nuclear experimentalists, etc. -- are limited to 50 mSv whole body dose in any one given year.

Table 4.2

Tabulation of selected dose equivalents

North American average from all sources	~ 2 mSv/y	(0.2 rem/y)
Terrestrial and cosmic radiation		
Banff, Canada	~ 1.5 mSv/y	(0.15 rem/y)
Kerola, India	~ 10 mSv/y	(1.0 rem/y)
Ramsar, Iran	~ 400 mSv/y	(40 rem/y)
Internal body doses		
Natural radioisotopes (^{40}K, ^{14}C, ...)	~ 0.3 mSv/y	(.03 rem/y)
Heavy smoking (^{210}Po, ...)	~ 80 mSv/y	(8 rem/y)

Discussion / Analysis 4.2

Undertake an analysis of your own radiation exposure in a typical year.

Constants	Exposure (mSv/y)
Cosmic and terrestrial	~ 0.8
Internal body	~ 0.3
Atmospheric fall-out	~ 0.05

Variables

Living/working above sea level (Add 0.02 for every 100 m) . . .	______
TV watching (Add 0.003 for every 3 hr/day . . .	______
Chest or dental x-ray (Add 0.1 for every occurrence . .	______
Flying (Add 0.001 for every hour aloft) . .	______
Housing (Add 0.06 if house contains masonry) .	______
Others (smoking, mountain climbing, etc) (Consult Health Physics text) .	______
TOTAL . . .	______

To think about: .

Is there, anywhere, a radiation-free environment?

Problems

4.1 Compute the shielding half-thickness of Al, W and U, for neutrons and gammas using the data of Table 4.1.

4.2 What is the probability that a 1 MeV gamma ray photon will pass through a 1 cm thick lead shield?

4.3 Given Eq. 4.7, what is the radiation flux a distance r from an infinitely long line source? (Consider the line source to consist of a continuum of point sources.)

4.4 It has been observed that μ_γ/ρ is approximately the same for all materials for gammas of a given energy. Why?

4.5 It is often stated that neutrons constitute nonionizing radiation. Examine this statement critically.

4.6 Examine possible means of incorporating multiple scattering of radiation in a shield in order to obtain an improved estimate of radiation penetration.

CHAPTER V

NEUTRON-NUCLEUS INTERACTIONS

The neutron has been found to be most effective in "rearranging" nuclear configurations and thus making possible important processes such as isotope production, fission reactions, and neutron multiplication. We consider here these and other related phenomena and establish some quantitative methods of analysis.

5.1 Basic Phenomena

We have shown that nuclear collisions generally involve a change in mass and therefore constitute exoergic or endoergic processes. One of the most exoergic nuclear reactions known is the neutron induced fission process which, for a most common case, is given by

$$n + {}^{235}U \rightarrow {}^{236}U^* \rightarrow \nu n + P_1 + P_2 . \tag{5.1}$$

Here, $\nu \simeq 2.3$, and P_1 and P_2 are fission products which vary in nuclear composition from one fission process to another; statistically, they are described by a distribution as suggested in Fig. 5.1. Here we also indicate the intermediate excited ${}^{236}U^*$ state to emphasize that two distinct processes occur: neutron absorption followed by fission.

The Q-value for this fission process is given by

$$\begin{aligned} Q_{fi} &= [(m_n + m_{235}) - (\nu m_n + m_1 + m_2)]c^2 \\ &\simeq 190\ \text{MeV}\ (\simeq 3\times10^{-11}\ \text{J}) . \end{aligned} \tag{5.2}$$

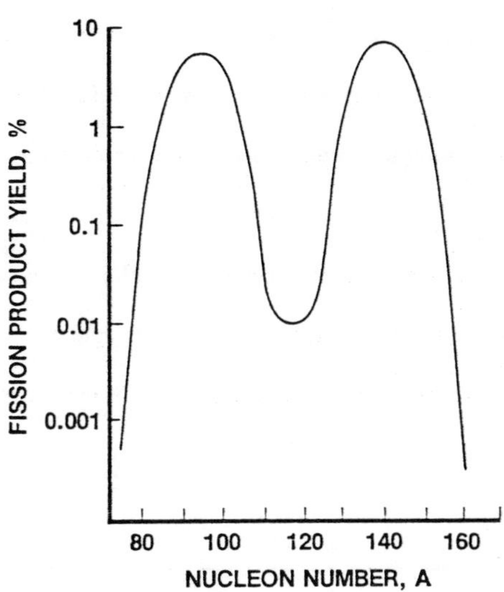

Fig. 5.1: Statistical distribution of fission products.

This energy thus released manifests itself in the following forms:

Kinetic energy of fission products:	~150MeV
Kinetic energy of emitted neutrons:	~5MeV
Energy of β^- and γ radiation:	~20MeV
Energy of neutrinos:	~15MeV
Fission Q-value:	190 MeV

All but the neutrino energy is, in principle, recoverable by a thermal energy conversion cycle; we will hereafter represent this component of Q_{fi} by Q^*_{fi}(i.e. $Q^*_{fi} \simeq 175$ MeV).

While the large energy release is one important consequence of a fission event, the observation that neutron multiplication also appears is of considerable significance because it makes possible the sustainment of a chain reaction. For example, if on average, exactly one of the fission neutrons fissions another fissile nucleus -- and the remaining $(\nu-1)$ neutrons are lost by escape from the medium or are absorbed by parasitic neutron absorption processes -- then a steady-state nuclear fission chain can be sustained. This chain is suggested in Fig. 5.2, where the initiating neutron is taken to be supplied from an external non-fission process such as a neutron emitting radioisotope; the reaction chain is terminated by the absorption of the chain carrier neutron in a neutron absorbing nucleus. We conclude therefore that at least three types of of nuclei are required for the operation of a fission reactor: a neutron source to initiate the reaction chain, fissile nuclei to sustain the chain, and neutron absorbing control materials to terminate or control the chain reaction.

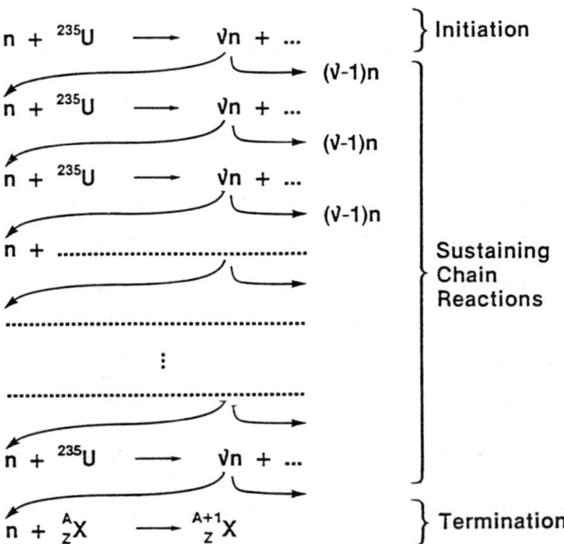

Fig. 5.2: Depiction of a neutron sustained fission chain reaction.

Each fission reaction releases Q^*_{fi} units of recoverable energy so that the thermal power density P_t is

$$P_t = R_{fi}\, Q^*_{fi}\,, \tag{5.3}$$

where R_{fi} is the number of fissions occurring per unit time in a unit volume -- that is the fission reaction rate density.

5.2 Neutron-Nucleus Processes

The neutron induced fissioning of a fissile nucleus is but one of several types of interactions which affect the population of neutrons in a fissioning media. We list and discuss several of the most relevant processes.

Elastic Scattering: This process is analogous to a billiard-ball type interaction in which kinetic energy is conserved. Using $^A{}_Z X$ as an arbitrary nucleus, a general elastic scattering event is represented by

$$n + {}^A_Z X \rightarrow n + {}^A_Z X\,. \tag{5.4}$$

Inelastic Scattering: In this case, the incident neutron possesses sufficient energy to excite the nucleus to an elevated nuclear energy level; we write the sequential processes as

$$n + {}^A_Z X \rightarrow (\;) \rightarrow n' + {}^A_Z X^*\,, \tag{5.5a}$$

and

$${}^A_Z X^* \rightarrow {}^A_Z X + \gamma\,. \tag{5.5b}$$

Here, n′ is to suggest that the neutron after the collision need not necessarily be the same; also, the asterisk in $^A{}_Z X^*$ is to indicate an excited state which subsequently decays by the emission of the de-excitation gamma ray photon(s).

Neutron Capture: In this very common process, the neutron becomes lodged in the nucleus, thus creating a new isotope; de-excitation gamma rays are invariably emitted almost concurrently:

$$n + {}^{A}_{Z}X \rightarrow (\ ) \rightarrow {}^{A+1}_{Z}X^{*} + \gamma . \tag{5.6}$$

Neutron-Induced Fission: For completeness, we list here again a particular fission process

$$n + {}^{A}_{Z}X \rightarrow (\ ) \rightarrow \nu n + {}^{A_1}_{Z_1}X^{*}_{1} + {}^{A_2}_{Z_2}X^{*}_{2}, \tag{5.7}$$

and show the nucleon balances as

$$A = (\nu - 1) + A_1 + A_2 , \tag{5.8a}$$

$$Z = Z_1 + Z_2 . \tag{5.8b}$$

The fission products are invariably in an excited state and decay by the emission of a variety of processes, but particularly beta decay,

$$ {}^{A_i}_{Z_i}X^{*}_{i} \rightarrow {}^{A_j}_{Z_j}Y_{j} + \ldots . \tag{5.9}$$

Indeed, the daughters of fission products may also be radioactive, yielding thereby a short radioactive chain.

Nucleon Emission: Nuclides also exist with the property of charged particle emission following neutron absorption; neutron multiplication may also occur. We suggest these processes in the following listing:

$$n + {}^{A}_{Z}X \rightarrow (\ ) \rightarrow \begin{cases} {}^{A}_{Z-1}X + p \\ {}^{A-3}_{Z-2}X + \alpha \\ {}^{A-1}_{Z}X + 2n \\ \cdot \\ \cdot \\ \cdot \end{cases} \tag{5.10}$$

The term scattering implies either or both elastic or inelastic scattering and all the remaining processes are designated neutron absorption. This is a useful distinction because the former processes do not affect the neutron population density.

The various possible neutron-nucleus interactions can be graphically depicted in the form of an event tree as follows:

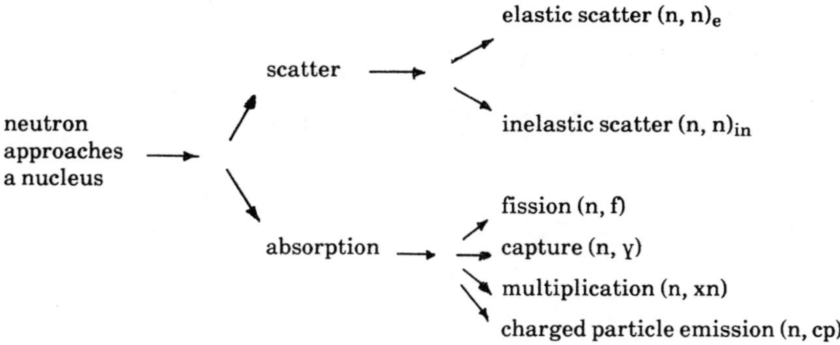

Each arrow represents an event to which a probability of occurrence can be assigned. This probability depends upon the nucleus involved and the kinetic energy of the incident neutron.

5.3 Neutron Cross Section

Reflecting upon the above discussion of the several types of neutron-nucleus interactions which can occur, as well as on the fission chain of Fig. 5.2 involving fission and non-fission processes, suggests the need for a comprehensive characterization of the various reactions and their rates of occurrence. For this purpose consider a medium which for now is taken to consist of only one type of nuclide -- for example pure aluminum ^{27}Al, pure ^{12}C, pure ^{16}O, etc. We take these atoms of density N_j all to be "stationary" about their equilibrium lattice position. Suppose further that a steady-state neutron population of density N_n is sustained in this medium by some external neutron source and -- if desired

-- also supplanted by possible neutron multiplication processes in the medium. The neutron population is taken to be in thermal equilibrium so that a well defined average neutron speed v can be defined. A typical fast-action "snapshot" of the neutron-nucleus interaction on a plane somewhere in the medium might appear something like that suggested in Fig. 5.3, below.

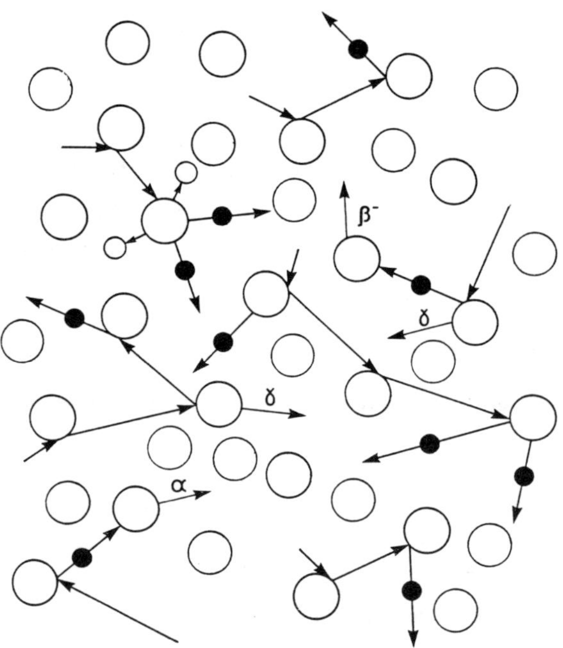

Fig. 5.3: "Short exposure image" of neutrons (black circles) interacting with the nuclei (open circles) in a medium.

The processes in Fig. 5.3 represent several types of interactions. Consider first only one particular type of neutron-nucleus interaction -- call it the r-type -- and define the reaction rate density for this type of process by the following:

$$R_r = \left(\begin{array}{c} \text{Number of r-type neutron-nucleus} \\ \text{interactions per unit volume per unit time} \end{array} \right). \tag{5.11}$$

We can expect that, if there is an increase in the neutron density N_n then there will be a corresponding increase in R_r; we also expect a similar trend if the "stationary" atom density N_j changes. Then, if a change in the average neutron speed, v, occurs, a corresponding change in the reaction rate will take place. We conclude therefore that a proportionality relationship of the form

$$R_r \propto v\, N_n\, N_j\,, \tag{5.12}$$

includes the important factors for the case of interest. Note that a zero for any of these three factors will, of course, imply a zero reaction rate density.

A proportionality relationship can always be cast into an exact equation by the introduction of a proportionality constant. This constant must reflect the details of the process which in our case involves r-type neutron-nucleus interactions of neutrons of speed v with nuclei of type j. We rewrite Eq. (5.12) to give

$$R_r = \sigma_r^j\, v\, N_n\, N_j\,, \tag{5.13}$$

where r and j associated with the constant σ_r^j are the appropriate identifiers applicable for neutrons of average speed v. This proportionality constant σ_j^r is called the microscopic cross section and, since all the other symbols possess specified units R_r [=] cm^{-3} s^{-1}, v [=] cm s^{-1}, N_n [=] cm^{-3}, N_j [=] cm^{-3} dimensional homogeneity requires that

$$\sigma_r^j\ [=]\ \text{cm}^2. \tag{5.14}$$

Hence, the microscopic cross section possesses units of area and may be interpreted as a reaction target area displayed by a nucleus of type j for a neutron of speed v to engage in an r-type neutron-nucleus interaction. This notion of a cross-sectional area, however, is not directly related to any geometric notions about the size of the nucleus. Appendix C provides another perspective on this cross section concept.

Typical microscopic cross sections are relatively small quantities; it has therefore been agreed to introduce the unit of "barn", abbreviated "b", such that

$$1\,b = 10^{-24}\,cm^2\ . \tag{5.15}$$

We list several neutron-nucleus cross sections in Table 5.1 for v = 2200 m/s; this corresponds to the most frequently encountered neutron speed for which the medium is at room temperature ($\simeq$ 300 K). Appendix A, Tables A.4 and A.5, provide a more extensive listing.

Table 5.1

Cross Sections of Selected Nuclides
(v = 2200 m/s)

Isotope	Cross Section (b)		
	σ_s	σ_f	σ_c
^{9}Be	7.0	0	0.01
^{10}B	2.1	0	3840
^{27}Al	1.4	0	2.41
^{59}Co	7.0	0	38.0
^{135}Xe	4.0	0	2.7×10^6
^{235}U	13.0	577	101
^{238}U	9.0	0	2.7

The following notation is frequently used for the commonly most important neutron-nucleus cross sections:

σ_s : scattering

σ_a: absorption

σ_f: fission

σ_c: capture

σ_{cp}: charged particle ejection

σ_t: total

Further, the following summation applies:

$$\begin{aligned}\sigma_t &= \sigma_s + \sigma_a \\ &= \sigma_s + \sigma_c + \sigma_f + \sigma_{cp} \\ &= \sum_j \sigma_j \ .\end{aligned} \tag{5.16}$$

Note that $\sigma_a = \sigma_c + \sigma_f + \sigma_{cp}$; such a summation of cross sections is possible because each represents a mutually exclusive event. Also scattering can be decomposed into elastic and inelastic scattering, e.g. $\sigma_s = \sigma_{s,e} + \sigma_{s,i}$

Before leaving the subject of reaction rate specification, we remove the idealization of a pure medium. Consider a medium to consist of j = 1, 2, 3, ... different nuclides, each of uniform density $N_1, N_2, N_3, \ldots$. For each, we have an associated microscopic cross section $\sigma_r^{(j)}$ for the r-type of neutron-nucleus process. By superposition, the reaction rate for the r-type of neutron-nucleus interaction is evidently

$$R_r = \sigma_r^{(1)} v N_n N_1 + \sigma_r^{(2)} v N_n N_2 + \sigma_r^{(3)} v N_n N_3 + \ldots = v N_n \left[\sum_{j=1} \sigma_r^{(j)} N_j \right] . \tag{5.17}$$

We now introduce two substitutions. In view of our discussion in Chapt. 4, Sec. 4.2, we write

$$\Phi_n = N_n v, \tag{5.18}$$

and call it the neutron flux. Further, we use

$$\begin{aligned}\sum_{j=1}^{J} \sigma_r^{(j)} N_j &= \sum_{j=1}^{J} \Sigma_r^j \\ &= \Sigma_r \ ,\end{aligned} \tag{5.19}$$

where Σ_r^j is called the macroscopic cross section for the j-type nuclide and Σ_r is the corresponding macroscopic cross section for the medium, involving r-type neutron-nucleus processes with the neutron kinetic properties defined by v.

Commonly encountered neutron densities seldom exceed $N_n \simeq 10^8$ cm^{-3}. The atom densities for single-nuclide materials are determined as discussed in Sec. 3.4. A more common situation is for several isotopes of an element to exist in a unit volume according to an atom ratio of

$$\gamma_j = \frac{N_j}{\sum_1^J N_j} . \tag{5.20}$$

In this case, then, the density of N_j is given by

$$N_j = \gamma_j \left(\frac{\rho_j}{M_j} \right) N_A , \tag{5.21}$$

where ρ_j and M_j are the elemental physical densities and the elemental gram molecular weights of the substance.

5.4 Interaction Distance

The depiction of neutron-nucleus collisions as suggested in Fig. 5.3 and the characterization of such interactions by a cross section, suggests that it should be possible to determine the average distance of neutron motion between collisions. This distance concept is useful for the development of a "feel" for the migrational properties of neutrons in various media and to allow the calculation of additional parameters.

In Chapter 4, we had considered the attenuation of a radiation beam as it entered a homogeneous medium. We found that the fraction of uncollided neutrons which have "survived" a distance x into the slab was given by

$$\frac{\phi_{c,n}(x)}{\phi_{c,n}(0)} = \exp(-\mu x) . \tag{5.22}$$

Here, $\phi_{c,n}(0)$ is the incident collimated neutron beam at $x=0$, $\phi_{c,n}(x)$ is the uncollided part of the collimated beam at x and μ is identified as the attenuation parameter.

Discussion / Analysis 5.1

Estimate the fuel burning rate of a 1000 MW fission reactor. (You may assume that only 235U is involved).

Core

$$\left(-\frac{\Delta m}{\Delta t}\right)_{fi} \propto P_{fi} = 10^3 \text{ MW}$$

$$\dot{M}_{fi} = R_{-f}\, m_f$$

m_f ← mass of a fissile nucleus, kg

R_{-f} ← fissile atom destruction rate, s⁻¹ (neutron absorption rate)

$$R_{-f} = R_f + R_c$$

$$= R_f\left(1 + \frac{R_c}{R_f}\right) = R_f\left(1 + \frac{\sigma_c^f N_f N_n v}{\sigma_f^f N_f N_n v}\right)$$

$$= R_f\left(1 + \frac{\sigma_c^f}{\sigma_f^f}\right) = R_f\left(1 + \frac{101}{577}\right)$$

Table A.4 (^{235}U)

$$= 1.18\, R_f$$

$$\therefore\ \dot{M}_{fi} = (1.18\, R_f)\, m_f$$

$$= 1.18\left(\frac{P_{fi}}{Q^*_{fi}}\right) m_f = 1.18 \times \frac{10^9 \text{ J/s}}{175 \text{ MeV} \times 1.6\times10^{-19} \text{ J/eV}} \times \frac{235 \text{ g/mole}}{0.6\times10^{24} \text{ atom/mole}}$$

$$= \boxed{1.5 \text{ kg/day}}$$

To think about:

How many tons of x% enriched uranium is needed for the above reactor in a year?

The meaning of Eq. (5.22) already tells us something of relevance in our attempt to determine the average distance of penetration until the neutrons interact in some way, that is, the distance until a neutron ceases being part of the uncollided beam component. To take this further requires a determination of μ in terms of the cross section σ_t^j for neutrons interacting in a medium of atom density N_j.

Recall that Eq. (5.22) was derived from the plausibility expression

$$-\frac{\Delta\phi_{c,n}}{\Delta x} \simeq \mu\,\phi_{c,n}\,, \tag{5.23}$$

where $\Delta\phi_{c,n}$ is the number of neutrons of speed v which have interacted while traversing a distance Δx, Fig. 4.6. That is, $\Delta\phi_{c,n}/\Delta x$ is the total number of neutron-nucleus interactions per cm^3-s at some coordinate x,

$$-\frac{\Delta\phi_{c,n}}{\Delta x} \simeq R_t(x)\,, \tag{5.24}$$

and is exact in the limit of $\Delta x \to 0$. According to Eq. (5.13), for neutrons of speed v in this collimated beam $N_{c,n}(x)$ in a medium of j-type nuclides of density N_j, we have

$$R_t(x) = \sigma_t^j\,N_j\,N_{c,n}(x)\,v\,. \tag{5.25}$$

Equating this relation with Eq. (5.23) yields a definition for the attenuation parameter μ for neutron-nucleus interactions:

$$\mu = \frac{\sigma_t^j\,N_j\,N_{c,n}(x)\,v}{\phi_{c,n}(x)}\,. \tag{5.26}$$

The useful identity is evidently

$$\phi_{c,n}(x) = N_{c,n}(x)\,v\,, \tag{5.27}$$

which, upon substitution in Eq. (5.26) gives for μ

$$\begin{aligned}\mu &= \sigma_t^j\,N_j\\ &= \Sigma_t^j\,,\end{aligned} \tag{5.28}$$

where Σ_t^j is the macroscopic total cross section for neutron interaction with j-type nuclides. We may therefore specifically write the uncollided neutron beam, Eq. (5.22), as

$$\frac{\phi_{c,n}(x)}{\phi_{c,n}(0)} = \exp(-\sigma_t^j N_j x)$$

$$= \exp(-\Sigma_t^j x) \quad . \tag{5.29}$$

This is a most useful expression and essential to the determination of an average distance of neutron motion to the next interaction. However, the exponential attenuation description makes it clear that some neutrons travel a short distance while others travel a long distance before interaction. Some probability consideration must now be introduced. Specifically, what we require is the probability that a neutron will travel without collision for a distance x and then interact; the average such distances will then be the mean-free-path of interest to us here.

A reinterpretation of Eq. (5.29) makes it clear that the probability of a neutron surviving without collision for a distance x given that it is present at $x=0$ is

$$Pr(x) = \frac{\phi_{c,n}(x)}{\phi_{c,n}(0)} = \frac{N_{c,n}(x)}{N_{c,n}(0)}$$

$$= \exp(-\Sigma_t^j x) \quad . \tag{5.30}$$

where the notation Pr(x) means the "probability of neutron survival for distance x". Then, the probability that a neutron interaction of any kind will take place in the following interval Δx is given by a rearrangement of Eq. (5.23) with the substitution for μ, Eq. (5.28):

$$Pr(\Delta x) = -\frac{\Delta\phi_{c,n}(x)}{\phi_{c,n}(0)} = -\frac{\Delta N_{c,n}}{N_{c,n}}$$

$$= \Sigma_t^j \Delta x \quad . \tag{5.31}$$

In the limit of $\Delta x \rightarrow dx$, we get

$$Pr(dx) = \Sigma_t^j dx \; , \tag{5.32}$$

for which

$$p(x)\,dx = \Sigma_t^j\,dx\,. \tag{5.33}$$

Here, p(x) is the evident probability density of neutron-nucleus interaction per unit distance.

The average distance of motion for any kind of interaction to occur -- known as the total mean-free-path -- now follows directly with the use of its mathematical definition

$$s_t = \frac{\int_0^\infty x\,p(x)\,dx}{\int_0^\infty p(x)\,dx}$$

$$= \int_0^\infty x\,\Sigma_t^j \exp(-\Sigma_t^j x)\,dx$$

$$= \frac{1}{\Sigma_t^j} \tag{5.34}$$

Thus, the macroscopic cross section can be related in a simple and direct manner to a physically meaningful dynamic property of a neutron in motion: namely, the average distance a neutron travels in a medium until something happens to it -- scattering, absorption, ... Note that this mean-free-path may be specialized to particular events $s_r = 1/\Sigma_r$ in a given medium.

5.5 Particle Dynamics

We enumerated in Section 4.2 the several important neutron-nucleus interaction processes of general interest in a fission energy context. As indicated, some reactions lead to neutron production while others lead to neutron removal; additionally, some new isotopes were produced while others were destroyed. For any of the particle densities, including neutrons, we may therefore write a dynamical equation of the form

Discussion / Analysis 5.2

Examine some neutron mean-free-paths for capture in fission reactor materials.

Have $s_c = \frac{1}{\Sigma_{c,i}} = \frac{1}{\sigma_c^{(i)} N_i}$

(↑ neutron mean-free-path for capture)

Consider two contrasting materials and their functions:

reactor control rod

(n) require short distance till capture

$\therefore s_c \rightarrow$ small, $\therefore \Sigma_c, \sigma_c \rightarrow$ large

reactor structural material

(n) prefer large distances till neutron capture (in order to minimize neutron losses and minimize radiation damage)

Data from Table A.5:

Cadmium (control rod material):

$$s_c = \frac{1}{\sigma_c N} = \boxed{8.8 \times 10^{-3} \text{ cm}}$$

Aluminum (structural material):

$$s_c = \frac{1}{\sigma_c N} = \boxed{76 \text{ cm}}$$

To think about:

Determine s_t in a two-medium region.

$$\frac{dN_j}{dt} = R_{+j} - R_{-j}, \tag{5.35}$$

where R_{+j} and R_{-j} are the reaction rate densities which add to or remove from the density of N_j. We shall consider some examples.

Consider first the case of neutron capture

$$n + {}^{A}_{Z}X \rightarrow {}^{A+1}_{Z}X + \gamma, \tag{5.36}$$

for which the neutron capture microscopic cross section is taken to be known as $\sigma^A{}_c$. The nuclide ${}^{A}{}_{Z}X$ of density N_A is evidently being destroyed and therefore, according to Eqs. (5.13) and (5.35) we write

$$\frac{dN_A}{dt} = 0 - \sigma_c^A v N_n N_A . \tag{5.37}$$

The production rate density of the product ${}^{A+1}X$ of density N_{A+1}, however, is equivalent to this destruction rate density and hence,

$$\frac{dN_{A+1}}{dt} = \sigma_c^A v N_n N_A - 0 . \tag{5.38}$$

As another example, the dynamical equation for fissile fuel nuclei of density N_f is written as

$$\frac{dN_f}{dt} = 0 - \sigma_a^f v N_n N_f , \tag{5.39}$$

where we emphasize that either neutron fission or neutron capture destroys a fissile nucleus; only scattering cross sections do not contribute to the transmutation of a nucleus.

The dynamical equation for the neutron density involves both production and destruction rate density terms

$$\frac{dN_n}{dt} = R_{+n} - R_{-n}$$

$$= \nu\sigma_f^f v N_n N_f - \sum_i \sigma_a^i v N_n N_i \tag{5.40}$$

Here, the first term identifies neutron multiplication by fission involving only fissile nuclides while the second term is the neutron removal rate density by all those nuclides which possess a non-zero absorption cross section. (With only a few exceptions, neutron multiplication by processes other than fission is generally negligible.) The dynamical equation for the neutron density, in this space-independent case, is more compactly written as

$$\begin{aligned}\frac{dN_n}{dt} &= \nu\sigma_f^f v N_n N_f - \sum_j \sigma_a^j v N_n N_j \\ &= [\nu\sigma_f^f N_f - \sum_j \sigma_a^i N_j]\; v N_n \\ &= [\nu\Sigma_f - \Sigma_a] v N_n . \end{aligned} \qquad (5.41)$$

Here, we have introduced the macroscopic cross section according to the definition of Eq. (4.18). We will have opportunity to use this equation in the next chapter.

Problems

5.1 Calculate the neutron absorption rate in pure ^{9}Be, ^{10}B, and ^{235}U. Use the data of Table 5.1 and take $N_n = 10^7$ neutrons/cm^3.

5.2 Determine the macroscopic cross sections Σ_s, Σ_f and Σ_c for ^{27}Al, ^{59}Co, ^{235}U, and ^{238}U using the microscopic cross section of Table 5.1.

5.3 Compute the neutron mean-free-path for scattering, fission and capture for the materials of Problem 5.2.

5.4 What material density and other conditions need to be satisfied in order for the neutron density in an arbitrary mixture to be at steady state?

CHAPTER VI

FISSION DYNAMICS

The preceding discussion of neutron-nucleus interactions provides the essentials for an introduction to fission reactor analysis. We first consider some dynamic aspects of fission energy and for this purpose assume a sufficiently large and homogeneous medium in which fission reactions and other neutron-nucleus processes occur.

6.1 Fission Power

Our medium of interest is taken to contain a uniform density of fissile nuclides, N_f, and others of various densities, N_1, N_2, N_3, ... The dominant exoergic reaction anywhere in this medium is of the fission reaction type

$$n + f \rightarrow (\ ) \rightarrow \nu n + P_1 + P_2 + Q_{fi} \ . \tag{6.1}$$

Here, n is a neutron of density N_n, f is a fissile nucleus -- generally ^{235}U -- of density N_f, ν is the average number of neutrons released per fission event -- typically close to 2.3; P_1 and P_2 are two fission products and Q_{fi} is the Q-value for a neutron induced fission and therefore the total nuclear energy released per fission event.

The rate at which fission events occur follows directly from Secs. 5.3 and 5.4 and is given by

$$R_{fi}(t) = \sigma_f^f \, v \, N_n(t) \, N_f(t) \, , \tag{6.2}$$

where σ_f^f is the microscopic fission cross section for neutrons of average speed v interacting with fissile nuclei f. Since $R_{fi}(t)$ possesses unit of fission reactions per cm^3-s and Q_{fi} possesses units of MeV per fission reaction, the recoverable fission power density is given by

$$P_{fi}(t) = R_{fi}(t)\, Q^*_{fi} = \sigma_f^f\, v\, N_n(t)\, N_f(t)\, Q^*_{fi} \,. \tag{6.3}$$

in units of MeV/cm^3-s and can be converted to J/cm^3-s or W/cm^3. Note, again, that Q^*_{fi} stands for the recoverable fission energy. Also, a small contribution from other nuclear energy-multiplication processes associated with secondary effects is not included here.

Our interest next is in any possible time variation in the various particle densities and in the reactor power; for that reason we have explicitly introduced the time variable t in Eqs. (6.2) and (6.3). The density of the fissile fuel, $N_f(t)$, will of course change with time because of fuel breeding and fuel burn-up. Indeed, the dynamical equation for $N_f(t)$ is simply

$$\frac{dN_f}{dt} = R_{+f} - \sigma_a^f\, v\, N_n(t)\, N_f(t) \,, \qquad N_f(0) = N_{f,o} \,. \tag{6.4}$$

Here, R_{+f} refers to conceivable nuclear fuel breeding processes and the second term describes fissile fuel destruction by neutron absorption. The initial condition listed here is referenced to an arbitrary $t = 0$.

The dynamical equation for the neutrons must include both neutron production by fission and neutron removal by absorption in whatever nuclei N_i exist in the medium with a non-zero absorption cross section $\sigma_a{}^i$. Hence, we obtain for $N_n(t)$

$$\frac{dN_n}{dt} = \nu\sigma_f^f\, v\, N_n(t)\, N_f(t) - \sum_i \sigma_a^i\, v\, N_n(t)\, N_i(t) \,, \qquad N_n(0) = N_{n,o} \,. \tag{6.5}$$

Here, the initial condition is also referenced to $t = 0$.

Examination of Eqs. (6.4) and (6.5) reveals a troublesome complication: the equations are coupled and nonlinear and, further, need to be augmented by rate equations for all the $N_i(t)$ particle densities listed in the last term of Eq. (6.5). An examination of the underlying nuclear processes,however, allows the introduction of some effective simplifications for the following reasons: the time scales for important changes in density of neutrons are of the order of seconds and minutes whereas the corresponding time scales for significant fractional changes in the nuclides are generally hours to months. We choose therefore to restrict our analysis for now to the shorter time intervals for which the atom densities change little

$$N_i(t) \rightarrow N_i, \quad i \neq n, \tag{6.6}$$

and write for the fission power density at time t, Eq. (6.3)

$$P_{fi}(t) = [\sigma_f^f \, v \, N_f \, Q^*_{fi}] \, N_n(t) . \tag{6.7}$$

Here all the factors in the brackets are constants; that is, the time variation in power is now dependent only upon time variations of the neutron densities.

Introducing the constancy of nuclides into Eq. (6.5) yields the neutron density rate equation more compactly

$$\begin{aligned} \frac{dN_n}{dt} &= \nu [\sigma_f^f \, N_f] \, v \, N_n(t) - \left\{ \sum_j \sigma_a^j N_j \right\} v \, N_n(t) \\ &= [\nu \Sigma_f - \Sigma_a] \, v \, N_n(t) \\ &= [\nu \Sigma_f / \Sigma_a - 1][\Sigma_a v] \, N_n(t) , \qquad N_n(0) = N_{a,o} . \end{aligned} \tag{6.8}$$

This simple first order differential equation is of the type

$$\frac{dx}{dt} = \omega x , \qquad x(0) = x_o , \tag{6.9}$$

with ω a parameter negative, zero, or positive. For time intervals for which ν, Σ_f and Σ_a may be considered sufficiently close to constant, integration leads to

$$N_n(t) = N_{n,o} \exp\{[(\nu \Sigma_f/\Sigma_a - 1)(\Sigma_a v)]t\} . \tag{6.10}$$

Upon substitution in the fission power density expression, Eq. (6.7), we obtain

$$P_{fi}(t) = [\sigma_f^f \, v \, N_f \, Q_{fi}] \, N_{n,o} \exp\{[(\nu \, \Sigma_t/\Sigma_a - 1)(\Sigma_a \, v)]t\}$$

$$= P_o \exp\{[(\nu \, \Sigma_f/\Sigma_a - 1)(\Sigma_a \, v)t]\} , \qquad (6.11)$$

where P_o is the initial fission power. Figure 6.1 provides a graphical depiction of the fission power depending upon the magnitude of $\nu \, \Sigma_f/\Sigma_a$ in relation to unity; $\nu \, \Sigma_f/\Sigma_a = 1$ has been assumed for $t < 0$.

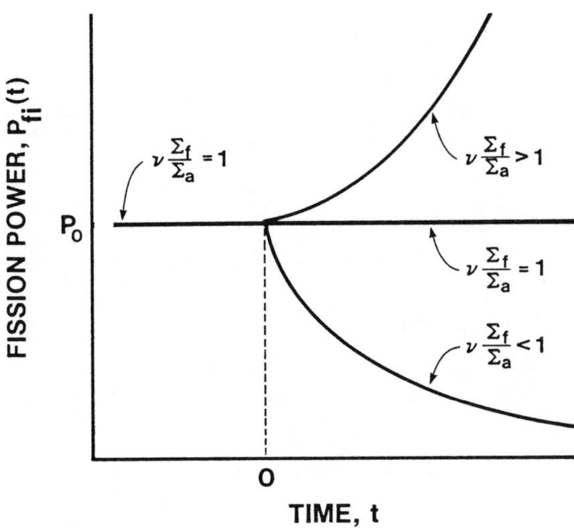

Fig. 6.1: Fission power as a function of time for three domains of $\nu \, \Sigma_f/\Sigma_a$.

Discussion / Analysis 6.1

By how much will a fission power increase in one second if a sudden control rod movement increases k_∞ from 1.000 to 1.001 ?

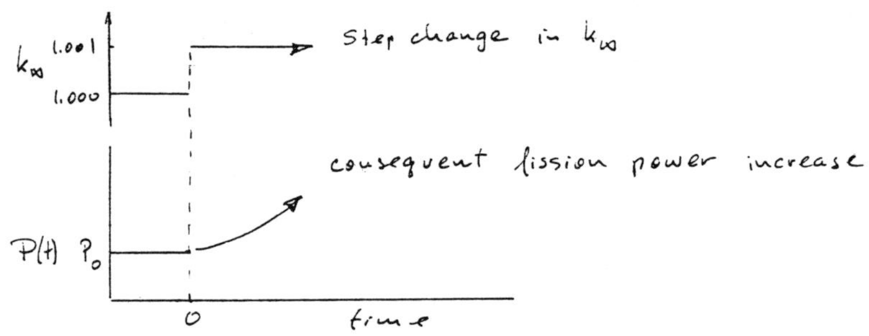

Have $\quad P_{fi}(t) = P_o \, e^{\left(\frac{k_\infty - 1}{\tau_n^*}\right)t}, \quad t > 0$

Based on other sources: choose $\tau_n^* = 0.1\text{ s}$

$$\therefore \; P_{fi}(1) = P_o \, e^{\left(\frac{1.000 - 1.001}{.1}\right)1} = 1.01 \, P_o$$

ie. 1% power rise in 1 s.

To think about:
what is the fastest possible way to shut down a reactor (i.e. scram condition)? For such a scram, how long until $P_{fi} \rightarrow 0.01 \, P_o$? (use $\tau_n^* = 0.1\text{ s}$)

6.2 Neutron Regulation

An examination of Eqs. (6.10) and (6.11) as well as Fig. 6.1 makes it evident that the magnitude of $\nu \Sigma_f/\Sigma_a$ relative to unity is the most important parameter which affects the form of fission power variation with time; the magnitude of $\Sigma_a v$ in Eq. (6.11) does not affect the functional form of the power transient. These results deserve further consideration.

Recall that

$$\nu \frac{\Sigma_f}{\Sigma_a} = \frac{\nu \sigma_f^f N_f}{\sum_j \sigma_a^j N_j}, \tag{6.12}$$

and hence this parameter is determined by the kind of nuclides in the medium, by their densities, and by their associated neutron-nucleus cross section for either fission in fissile isotopes or absorption in all isotopes. Evidently, changes in the density of these other nuclei may be an effective way of determining whether the power increases with time, is steady, or decreases.

Consider therefore a control rod composed of some material with a large absorption cross section. Inserting this rod further into the medium increases the overall density of neutron absorbing materials and hence, $\Sigma_a \nearrow$. This implies that

$$\nu \frac{\Sigma_f}{\Sigma_a} \searrow \quad \therefore P_{fi}(t) \searrow . \tag{6.13}$$

On the other hand, control rod withdrawal decreases the overall density of neutron absorbing materials and therefore $\Sigma_a \searrow$ so that

$$\nu \frac{\Sigma_f}{\Sigma_a} \nearrow \quad \therefore P_{fi}(t) \nearrow . \tag{6.14}$$

At some intermediate rod position, we will have

$$\nu \frac{\Sigma_f}{\Sigma_a} = 1 \quad \therefore P(t) \rightarrow . \tag{6.15}$$

We depict in Figure 6.2 how control insertions-withdrawals can change the power density of a fissioning medium with no independent neutron sources.

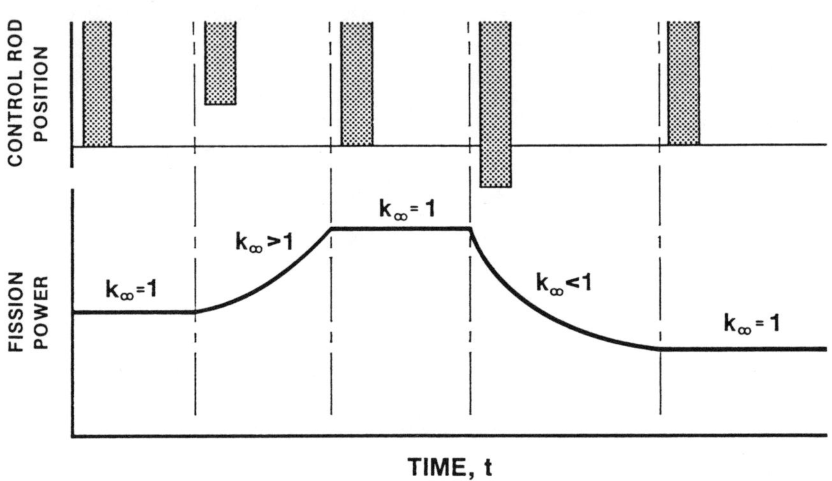

Fig. 6.2: Graphical relationship between control rod position, power production, and the neutron multiplication parameter.

The above suggest thats we define, for this space-independent reactor, a parameter

$$k_\infty = \nu \frac{\Sigma_f}{\Sigma_a} = \frac{\nu\Sigma_f N_n v}{\Sigma_a N_n v} = \frac{R_{+n}}{R_{-n}} , \tag{6.16}$$

and call it the neutron multiplication since its origin rests in the ratio of neutron production rate divided by the neutron absorption rate. Its range can be used to define a reactor state by the following designations:

$$k_\infty = \begin{cases} > 1 \text{, supercritical state} \\ = 1 \text{, critical state} \\ < 1 \text{, subcritical state} \end{cases} \tag{6.17}$$

The symbol ∞ is here listed to indicate that in the original defining equation for the neutron density, Eq. (6.8), no neutron leakage has been included; in other words, an infinite size reactor is assumed.

The product $\Sigma_a v$ in Eqs. (6.10) and (6.11) requires careful interpretation at several levels of relevance. Note first that this term is always positive and hence it does not affect the functional form of $N_n(t)$ or $P_{fi}(t)$; it does, however, affect the rate of change of these time dependent functions for $\nu\Sigma_f/\Sigma_a \neq 1$.

The product $\Sigma_a v$ may be interpreted as follows. We take the mean-free-path of neutron migration until absorption to be $s_a = 1/\Sigma_a$, Eq. (5.34), for neutrons moving at an average speed v. This average distance is traversed by neutrons during a time τ_n such that

$$s_a = v\,\tau_n \, . \tag{6.18a}$$

Therefore, we define a characteristic mean neutron lifetime τ_n for neutrons of speed v by

$$\tau_n = \frac{s_a}{v} = \frac{1}{\Sigma_a v} \, . \tag{6.18b}$$

The most important consideration in determining when $\tau_n = 1/\Sigma_a v$ can be used in the dynamic description of an actual fission reactor rests upon the extent to which the assumptions of the model used apply adequately to a fission reactor. Here, then we must recognize the following:

1. Rather than having all neutrons possess the same speed v, actual reactors contain neutrons of a very large range of speeds.
2. Rather than having all neutrons appear only as a result of fission reactions, a small but significant fraction of neutrons also appear as the result of the decay of radioactive fission products.

Both of these considerations lead to the recognition that a more appropriate neutron lifetime should be τ^*_n which relates to $\tau_n = 1/\Sigma_a v$ by

$$\tau^*_n = \tau_n + fc\left(\begin{array}{c}\text{neutron slowing down time and} \\ \text{constants of delayed neutron precursor}\end{array}\right) \tag{6.18c}$$

This suggests that the neutron and power variations, Eqs. (6.10) and (6.11), with time are simply

$$N_n(t) = N_{n,o} \exp\left\{\left(\frac{k_\infty - 1}{\tau^*_n}\right)t\right\} , \tag{6.19a}$$

and

$$P_{fi}(t) = P_o \exp\left\{\left(\frac{k_\infty - 1}{\tau^*_n}\right)t\right\} . \tag{6.19b}$$

for k_∞ and $\tau_n{}^*$ as constant. Exponential time variations, with k_∞ as a bifurcation parameter, thus characterize the dynamics of a fission reactor in a first approximation.

6.3 Fuel Burning

Reactor control, as we indicated, involves relatively short time periods. Our interest now is to examine time variations of fissile fuels which involve much longer time scales. In contrast to Eq. (6.6), we now take the neutron density to be constant,

$$N_n(t) \simeq N_n , \tag{6.20}$$

which may be accomplished by continuous control rod repositioning in order to sustain, on average, $k_\infty = 1$. The fission power density associated with a constant neutron population and time dependent fuel density, $N_f(t)$, is therefore given by

$$P_{fi}(t) = [\sigma_f{}^f v \, Q^*_{fi} \, N_n] \, N_f(t) . \tag{6.21}$$

The rate at which the nuclear fuel diminishes in the reaction medium is given -- for the case of no fissile fuel breeding --by the dynamical equation for N_f:

$$\frac{dN_f}{dt} = R_{+f} - R_{-f}$$

$$= 0 - [\sigma_c^f N_f(t) N_n v + \sigma_f^f N_{fi}(t) N_n v]$$

$$= - [\sigma_a^f v N_{n,o}] N_f(t) , \qquad N_f(0) = N_{f,o} . \tag{6.22}$$

The connection between the total recoverable energy extractable from an initial natural supply of fissile fuel with no breeding, follows from the recognition that P(t) in Eq. (6.21) is the rate of energy release so that

$$\frac{dE_{fi}}{dt} = \{\sigma_f^f v Q^*_{fi} N_{n,o}\} N_f(t) . \tag{6.23}$$

Division of this equation by Eq. (6.22) yields the fission energy recoverable per fissile nucleus destroyed

$$\frac{dE_{fi}/dt}{dN_f/dt} = \frac{dE_{fi}}{dN_f} = - \frac{[\sigma_f^f v Q^*_{fi} N_{n.o}] N_f(t)}{[\sigma_a^f v N_{n,o}] N_f(t)} . \tag{6.24}$$

That is,

$$\frac{dE_{fi}}{dN_f} = - \frac{\sigma_f^f}{\sigma_a^f} Q^*_{fi} , \tag{6.25}$$

and hence, for an initial inventory of fissile fuel of $N_{f,o}$, the total recoverable energy released follows by integration

$$\int_0^{E_{fi,tot}} dE = - \int_{N_{f,o}}^{0} \frac{\sigma_f^f}{\sigma_a^f} Q^*_{fi} dN_f$$

$$= - \frac{\sigma_f^f}{\sigma_a^f} Q^*_{fi} \int_{N_{f,o}}^{0} dN_f , \tag{6.26}$$

which yields

$$E_{fi,tot} = \left\{ \frac{\sigma_f^f}{\sigma_a^f} Q^*_{fi} \right\} N_{f,o} . \tag{6.27}$$

An evaluation of the only known naturally occurring fissile fuel, ^{235}U, together with the known σ_f^f, σ_a^f and Q^*_{fi} as natural constants had long ago revealed that the total energy thus extractable would be but a small fraction of global energy requirements even in the intermediate term. It was, however, soon found that although other thermally fissile nuclei do not exist, there was available a significant supply of fertile fuel which could be transmuted into fissile fuel by nuclear breeding based on neutron capture followed by radioactive decay. Two such breeding chains are known:

$$n + {}^{232}Th \rightarrow {}^{233}Th \underset{(22\ m)}{\rightarrow} {}^{233}Pa \underset{(27\ d)}{\rightarrow} {}^{233}U \tag{6.28a}$$

$$n + {}^{238}U \rightarrow {}^{239}U \underset{(23\ m)}{\rightarrow} {}^{239}Np \underset{(2.3\ d)}{\rightarrow} {}^{239}Pu \tag{6.28b}$$

The numbers in brackets represent the half-lives of the corresponding radionuclides. The end product ^{233}U and ^{239}Pu are fissile nuclei similar to ^{235}U.

It is estimated that ^{232}Th is about 500 times as plentiful as ^{235}U and ^{238}U occurs 138 times as frequently as ^{235}U; together, the fertile nuclei could -- in the absence of other limitations -- extend the nuclear fuel supply considerably with the bred fissile fuels ^{233}U and ^{239}Pu. The immediate restriction is the availability of neutrons to initiate the above breeding chains.

The current mainline approach to breeding of fissile fuel by the breeding chains involves the placement of ^{232}Th or ^{238}U into fission reactors and allowing some of the remaining $(\nu - 1)$ fission neutrons which are not needed to sustain the fission chain to be captured by these fertile nuclei to sustain the breeding chains according to Eq. (6.28). The bred fissile nuclei, ^{233}U or ^{239}Pu, would then be used in fission reactors just like the initially supplied ^{235}U.

6.4 Breeding and Energy Extension

The preceding discussion makes it obvious that the breeding of fissile fuel can substantially extend the total energy extractable by fission processes. We consider this assessment by introducing a particular reactor parameter.

Consider a reaction domain involving, among other neutron nucleus reactions, the concurrent fissioning of fissile nuclei and the breeding of fissile nuclei by neutron capture in fertile atoms according to either Eq. (6.28a) or Eq. (6.28b). Let f again be an arbitrary fissile nucleus (e.g. ^{235}U) and g an arbitrary fertile nucleus (^{232}Th or ^{238}U). The burning and breeding reactions of interest may be represented by

$$n + f \rightarrow (\ ) \rightarrow \nu n + P_1 + P_2 + Q_{fi}\,, \tag{6.29a}$$

and

$$n + g \rightarrow \quad \ldots \quad \rightarrow f\,. \tag{6.29b}$$

Here, in the second equation, we have compressed the two intermediate beta decay processes; this is fully justified if we limit ourselves to equilibrium fuel cycle conditions which involves time scales of the order of months.

We amplify the couplings between these two equations by indicating the linkages between the appropriate nuclear species:

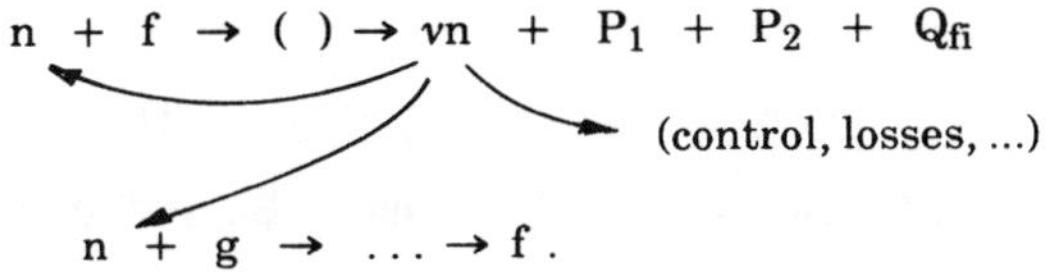

$$\begin{array}{l} n + f \rightarrow (\ ) \rightarrow \nu n + P_1 + P_2 + Q_{fi} \\ \qquad\qquad\qquad\qquad \searrow \text{(control, losses, ...)} \\ n + g \rightarrow \ldots \rightarrow f\,. \end{array} \tag{6.30}$$

This represents a kind of symbiosis in which a fission reaction provides neutrons for a breeding reaction which then yields fuel to continue the fission process. These two reactions proceed at the following rates:

$$R_{fi} = \sigma_f^f\, v\, N_n(t)\, N_f(t)\,, \tag{6.31a}$$

$$R_{br} = \sigma_c^g\, v\, N_n(t)\, N_g(t)\,. \tag{6.31b}$$

We now introduce a parameter C to be called the conversion ratio, defined by

$$C = \frac{\text{production rate of fissile nuclei}}{\text{destruction rate of fissile nuclei}} > 0\,, \tag{6.32}$$

and which, in terms of specific reaction rates, is given as

$$C = \frac{\sigma_c^g\, v\, N_n(t)\, N_g(t)}{\sigma_a^f\, v\, N_n(t)\, N_f(t)} = \frac{\sigma_c^g\, N_g(t)}{\sigma_a^f\, N_f(t)}\,. \tag{6.33}$$

A self-evident classification of fission reactors may then be established depending upon the magnitude of C:

$$C = \begin{cases} \simeq 0: & \text{Burner Reactor} \\ < 1: & \text{Converter Reactor} \\ = 1: & \text{Self-Sufficient Reactor} \\ > 1: & \text{Breeder Reactor}\,. \end{cases} \tag{6.34}$$

Typical values of C for existing commercial fission reactors are in the range of 0.3 to 0.6.

With the fissile fuel replenishment capacity thus defined, we can assess the importance of the breeding reaction, Eq. (6.29b), by an examination of the dynamical equation for $N_f(t)$ in a fission reactor core,

$$\frac{dN_f}{dt} = R_{+f} - R_{-f}$$

$$= \sigma_c^g\, v\, N_n(t) N_g(t) - \sigma_a^f\, v\, N_n(t)\, N_f(t) \tag{6.35}$$

Here, the possibility of having the breeding chain, Eq. (6.28), disrupted by neutron absorption is taken to be sufficiently small. Then, using Eq. (6.33)

$$\frac{dN_f}{dt} = C\,\sigma_a^f\, v\, N_n(t)\, N_f(t) - \sigma_a^f\, v\, N_n(t)\, N_f(t)$$

$$= (C-1)\,\sigma_a^f\, v\, N_n(t)\, N_f(t) \tag{6.36}$$

This equation constitutes a general dynamical description for the fissile fuel evolution in a fissioning domain. As indicated, however, it depends not only upon the time variation of the conversion ratio C due to its dependence on $N_g(t)$ and $N_f(t)$, Eq. (6.33). These complications notwithstanding, an analysis of Eq. (6.36) is still possible, which is useful in describing the fissile fuel inventory in a fission reactor.

As discussed in Sec. 6.2, the neutron density can be varied within specified bounds by control rod motion. When these bounds are reached, the reactor must be refueled. We choose therefore to restrict our analysis to the case between refueling for which an average N_n can be assumed.

According to Eq. (6.33) the conversion ratio is zero if no breeding materials N_g are present in the reactor. Then, for the case of a converter reactor, $0 < C < 1$, and according to the reaction linkage of Eq. (6.30) as well as Eq. (6.35), both $N_g(t)$ and $N_f(t)$ decrease with time so that C can be considered to be approximately constant. For the case of a self-sufficient reactor, a constant $C \simeq 1$ is, in principle, possible only if a replenishment of $N_g(t)$ equal to its destruction rate is maintained; this can seldom be done so that a constant $C = 1$ is a weak assumption in our model. Even weaker is the case of sustaining $C > 1$ as a parameter for a breeder reactor.

Nevertheless, many useful reactor analyses can be undertaken with the assumption that the conversion ratio varies only within acceptable bounds. The time variation of the fissile fuel in the fissioning domain varies for the several reactor types and over appropriate time intervals has the form indicated in Fig. 6.3.

Further thought suggests that the case of reactor refueling can be described by a saw-tooth pattern in which the trends of Fig. 6.3 are repeated within narrow bounds. In practice, however, another complication arises because most reactors are only partially refueled at regular intervals.

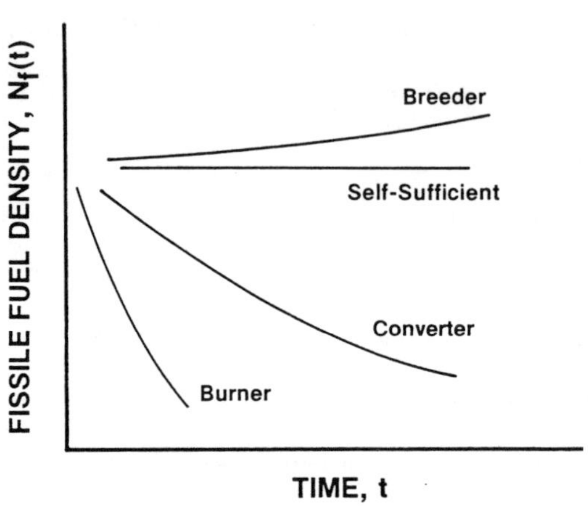

Fig. 6.3: Illustration showing the effect of nuclear breeding on the extension of fissile fuel.

Problems

6.1 What limitations on ν, Σ_f and Σ_a would need to be imposed if the fission power increase in 60 s is not to exceed 2%?

6.2 The power density of a thermal fission reactor is of the order of 10 MW/m^3. For $N_n \simeq 10^7$ cm^{-3} and the data of Table 5.1, determine the average ^{235}U density.

6.3 Based on the information herein, calculate $E_{tot}/N_{f,o}$ of Eq. (6.27).

6.4 Explore some solution possibilities of both Eq. (6.4) and Eq. (6.5) without the imposition of constant particle densities for N_n and N_f.

Discussion / Analysis 6.2

Consider a converter fission reactor (ie $c < 1$) which requires refueling when a fraction x of its initial fuel is burned. How does the magnitude of c affect the burn time until refueling?

typical unit core volume containing $N_f(t)$

$$\frac{dN_f}{dt} = R_{+f} - R_{-f}$$

$$= R_{-f}\left(\frac{R_{+f}}{R_{-f}} - 1\right)$$

$$= \sigma_a^f N_f N_n v\,(c - 1)$$

$$= \sigma_a^f N_n v\,(c-1)\,N_f(t)$$

take as constant K

Integrate

$$N_f(t) = N_f(0)\,e^{K(c-1)t} \qquad t > 0,\ c < 1$$

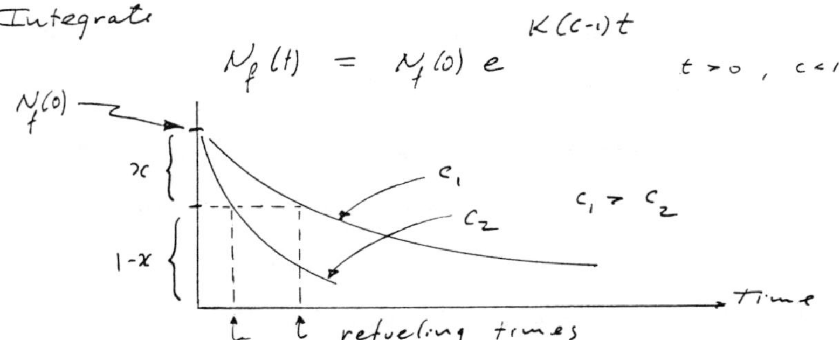

To think about:

Develop analytical expressions for increased fuel resource utilization with increasing conversion ratio c.

CHAPTER VII

FISSION REACTORS

A collection of fissile and other nuclides may or may not sustain a steady-state fission reaction chain. This important reactor characteristic depends upon the geometry of the assembly and its material composition. Our aim in this chapter is to determine the relationship between these three important parameters: criticality, composition, geometry.

7.1 Finite Geometry

We extend our discussion of the infinite fission reactor of Chapt. VI to a finite assembly containing both fissile and other neutron absorbing materials. The finiteness of the medium in which fission reactions occur has important implications for the neutron population because of neutron leakage. We will find that significant spatial dependencies will emerge and need to be incorporated in the description. The simplest and most practical geometries are spherical, rectangular, and cylindrical, Fig. 7.1, each characterized by an easily described non-reentrant surface. These assemblies are taken to be totally unshielded so that neutrons leaking out will not be reflected back.

Actual reactors generally possess one of the three geometries of Fig. 7.1 and contain several kinds of materials which serve different purposes:

(a) fuel: fissile ^{235}U and fertile ^{238}U generally in the form of the oxide UO_2;

(b) fuel cladding: aluminum, stainless steel or a zirconium alloy encases the UO_2 to contain radioactive fission products;

(c) coolant: either liquid H_2O or D_2O or a gas such as helium or carbon dioxide, and flowing through the reactor to transport the nuclear heat generated;

(d) control rods: invariably an alloy containing cadmium, boron or gadolinium and/or other neutron absorbing elements;

(e) moderator: liquid or solid which serves to slow down the high energy neutrons in order to increase the fission rate. (This applies to thermal reactor only; fast breeder reactors must minimize such materials).

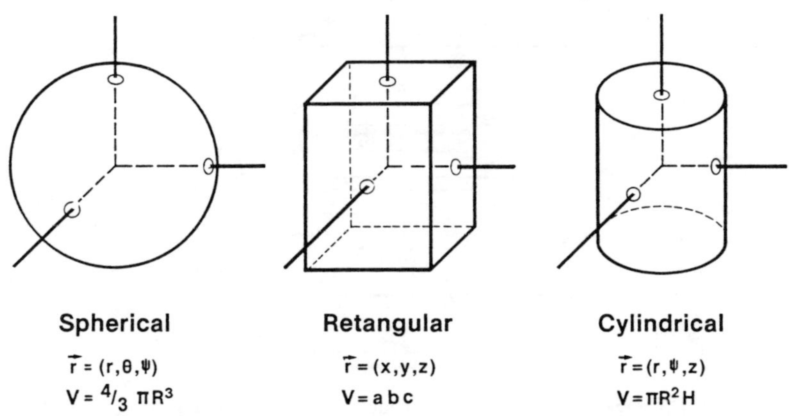

Fig. 7.1: Non-reentrant geometries of general interest for the assessment of fission criticality.

These four material groups, together with structural components and in-core instrumentation, are assembled into a suitable lattice configuration to form a reactor core. Figure 7.2 depicts a planar representation of such an arrangement and evidently constitutes a very heterogeneous assembly. The spatial variation in nuclear power generated will be considerable and therefore the total power is given by integration over the entire reactor volume:

$$P_t(t) = \int_V P_{fi}(\mathbf{r}, t)\, dV$$

$$= \int_V \sigma_f^f \, v \, Q^*_{fi} \, N_f(\mathbf{r}, t) \, N_n(\mathbf{r}, t) \; dV \, . \tag{7.1}$$

Here, σ_f^f, v and $\mathbf{Q^*}_{fi}$ are constant but the fuel density, $N_f(\mathbf{r}, t)$, and the neutron density, $N_n(\mathbf{r}, t)$, will in general vary with space and time, $(\mathbf{r}, t)$.

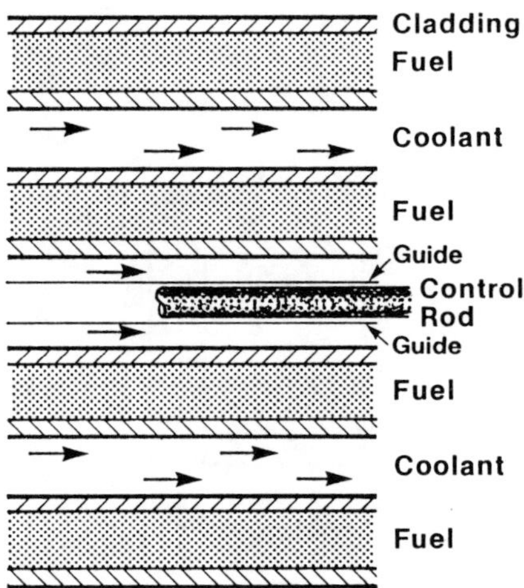

Fig. 7.2: Planar section of fission reactor core illustrating how fuel, cladding, coolant and control rods may be assembled. The coolant is here assumed to be a liquid so it also serves as a neutron moderator.

The evaluation of Eq. (7.1) in the form given is a very cumbersome process. However, for our purposes of criticality assessment, we may assume that all nuclides are considered to be uniformly distributed throughout the reactor core. That is, we will take

$$N_i(\mathbf{r}, t) \rightarrow N_i(t), \quad i \neq n, \tag{7.2}$$

for all particle densities except for the neutrons. Further, in our criticality analysis here, we will restrict ourselves only to relatively short time intervals for which all the nuclide densities do not change significantly; hence, all the materials can further be taken to be constant with time,

$$N_i(t) \rightarrow N_i, \quad i \neq n, \tag{7.3}$$

so that the recoverable thermal power is now given as

$$P_t(t) = \sigma_f^f \, v \, Q^*_{fi} \, N_f \int_V N_n(\mathbf{r}, t) \, dV. \tag{7.4}$$

The determination of the space-time variation of the neutron density, $N_n(\mathbf{r}, t)$, within the context of a criticality assessment, is now our main interest.

In contrast to our discussion of the infinite reactor, we must now incorporate the process of neutron leakage. Not only does this affect the total number of neutrons in the assembly but it introduces a significant spatial variation because neutrons near the enclosing surface are more likely to escape than those in the more central region.

We take a suitably small volume element ΔV enclosed by a surface of area ΔA and located at an arbitrary point in the reactor assembly of interest, Fig. 7.3. The total number of neutrons in this volume element satisfies the neutron balance dynamical equation which accounts for all processes that affect the neutron density:

$$\frac{\partial}{\partial t} \int_{\Delta V} N_n(\mathbf{r}, t) \, dV = \int_V \left\{ (R_{+n})_{fission} - (R_{-n})_{absorption} - (R_{-n})_{leakage} \right\} dV$$

$$= \int_{\Delta V} \left\{ \left(\nu \, \sigma_f^f \, v \, N_f N_n(\mathbf{r}, t) - \sum_i \sigma_a^i \, v \, N_i \, N_n(\mathbf{r}, t) - (R_{-n})_{leakage} \right) \right\} dV. \tag{7.5}$$

The last term, that is the neutron leakage rate component, is the new and important term and requires closer examination.

7.2 Neutron Leakage

In order to obtain compact and useful expressions for the neutron leakage rate we take a closer look at the flow of neutrons across some surface ΔA enclosing the arbitrary volume element ΔV, Fig. 7.3. The important concept is the need for a vector quantity of the neutron flow relative to the normal of the surface ΔA. We introduce the neutron current $\mathbf{J}_n(\mathbf{r}, t)$ at the point $\mathbf{r}$ and at time t by vectorial addition of all the neutrons passing through a differential area dA according to the following

$$\mathbf{J}_n(\mathbf{r}, t) \cdot \mathbf{\Omega}\, dA = \begin{pmatrix} \text{Net number of neutrons which pass} \\ \text{through area dA in a unit of time} \end{pmatrix} . \tag{7.6}$$

Here $\mathbf{\Omega}$ is a unit vector normal to dA. The net leakage of neutrons for ΔV is evidently the normal component of $\mathbf{J}_n(\mathbf{r}, t)$ integrated over its enclosing surface ΔA; that is,

$$\int_{\Delta V} (R_{-n})_{leakage}\, dV = \int_{\Delta A} \mathbf{J}_n(\mathbf{r}, t) \cdot \mathbf{\Omega}\, dA . \tag{7.7}$$

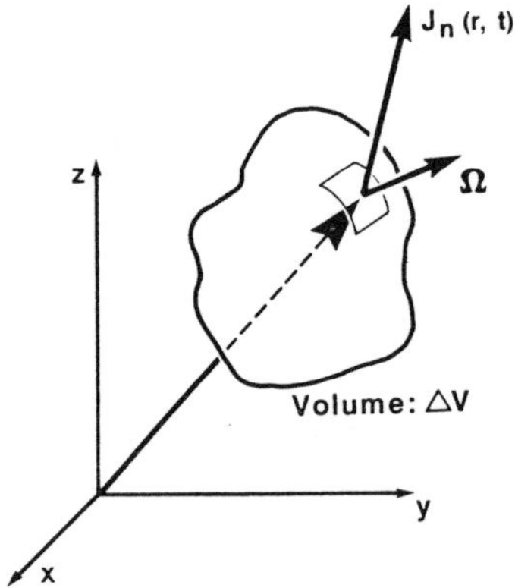

Fig. 7.3: Relationship between the neutron current vector $\mathbf{J}_n(\mathbf{r},t)$ and unit vector $\mathbf{\Omega}$ for a volume element ΔV enclosed by a non-reentrant surface ΔA.

According to Gauss' Divergence Theorem, a surface integration can be related to a volume integration which, for our case, yields

$$\int_{\Delta A} \mathbf{J}_n(\mathbf{r}, t) \cdot \mathbf{\Omega}\, dA = \int_{\Delta V} \nabla \cdot \mathbf{J}_n(\mathbf{r}, t)\, dV\,, \tag{7.8}$$

with ∇ as the del-operator. The dynamical equation for the neutron density, Eq. (7.5), may therefore be written as

$$\frac{\partial}{\partial t} \int_{\Delta V} N_n(\mathbf{r}, t) = \int_{\Delta V} \nu \Sigma_f\, v\, N_n(\mathbf{r}, t)\, dV - \int_{\Delta V} \Sigma_a\, v\, N_n(\mathbf{r}, t)\, dV - \int_{\Delta V} \nabla \cdot \mathbf{J}_n(\mathbf{r}, t)\, dV\,, \tag{7.9}$$

where we have used the definitions

$$\Sigma_f = \sigma_f^f N_f\,, \tag{7.10a}$$

and

$$\Sigma_a = \sum_i \sigma_a^i N_i\,, \tag{7.10b}$$

for our homogeneous medium.

Further, Eq. (7.9) may also be written as

$$\int_{\Delta V} \left\{ \frac{\partial}{\partial t} N_n(\mathbf{r}, t) - \nu \Sigma_f\, v\, N_n(\mathbf{r}, t) + \Sigma_a\, v\, N_n(\mathbf{r}, t) + \nabla \cdot \mathbf{J}_n(\mathbf{r}, t) \right\} dV = 0\,, \tag{7.11}$$

and, since ΔV is arbitrary, the integrand must be identically zero. Hence, we obtain the following dynamical equation for the neutron density, $N_n(\mathbf{r}, t)$, -- a scalar quantity -- and for the neutron current, $\mathbf{J}_n(\mathbf{r}, t)$ -- a vector quantity:

$$\frac{\partial}{\partial t} N_n(\mathbf{r}, t) = \nu \Sigma_f\, v\, N_n(\mathbf{r}, t) - \Sigma_a\, v\, N_n(\mathbf{r}, t) - \nabla \cdot \mathbf{J}_n(\mathbf{r}, t)\,. \tag{7.12}$$

This, however, is but one equation which contains two unknown functions, $N_n(\mathbf{r}, t)$ and $\mathbf{J}_n(\mathbf{r}, t)$; another relationship connecting these functions is obviously required.

It has frequently been observed that spatial variations of a scalar quantity induce a vector quantity. For example, in mass transfer it is known that a particle current from high to low particle densities always takes place; similarly, in heat transfer a heat current from a region of high temperature to one of low temperature is known to occur. Such processes are typically characterized by a proportionality relationship of the type

$$\mathbf{q}(\mathbf{r}) \propto -\nabla\phi(\mathbf{r}) , \tag{7.13}$$

where $\phi(\mathbf{r})$ is a scalar and $\mathbf{q}(\mathbf{r})$ a vector. The del-operation generates a gradient of $\phi(\mathbf{r})$ and the minus sign indicates that the direction of $\mathbf{q}(\mathbf{r})$ is from higher values of $\phi(\mathbf{r})$ towards lower values. The proportionality of Eq. (7.13) may be transformed into an equation by the introduction of a constant which is called a diffusion coefficient appropriate to the phenomenon and the medium of interest.

For many cases of neutron migration, it is known that the neutron current $\mathbf{J}_n(\mathbf{r}, t)$ is related to the neutron flux $\phi_n(r,t) = v\, N_n(\mathbf{r}, t)$ by a relationship of the form of Eq. (7.13),

$$\mathbf{J}_n(\mathbf{r}, t) \propto -\nabla\{v\, N_n(\mathbf{r}, t)\} , \tag{7.14a}$$

or, more specifically,

$$\begin{aligned} \mathbf{J}_n(\mathbf{r}, t) &= -D\,\nabla\{v\, N_n(\mathbf{r}, t)\} \\ &= -D\,\nabla\,\phi_n(\mathbf{r},t) \end{aligned} \tag{7.14b}$$

with D as the neutron diffusion coefficient for the medium of interest and for neutron migration of average speed v. The name Fick's Law of Diffusion is often assigned to this relationship although, because of its approximation, the designation 'Fick's rule' of relating the scalar neutron flux to the vector neutron current seems more appropriate.

Equation (7.14b) provides the additional relationship in order to eliminate $\mathbf{J}_n(\mathbf{r}, t)$ in Eq. (7.12). Substituting, we obtain

$$\begin{aligned} \frac{\partial}{\partial t} N_n(\mathbf{r}, t) &= \nu\Sigma_f\, v\, N_n(\mathbf{r}, t) - \Sigma_a\, v\, N_n(\mathbf{r}, t) - \nabla\cdot[-D\nabla\{v\, N_n(\mathbf{r}, t)\}] \\ &= \nu\Sigma_f\, v\, N_n(\mathbf{r}, t) - \Sigma_a\, v\, N_n(\mathbf{r}, t) + D\, v\, \nabla^2 N_n(\mathbf{r}, t) . \end{aligned} \tag{7.15a}$$

Here, in the last term, we have taken D as independent of spatial coordinates and used $\nabla^2 = \nabla\cdot\nabla$ as the Laplacian operator.

Recalling that $\phi_n(\mathbf{r}, t) = v\, N_n(\mathbf{r}, t)$ with v as a constant, leads to two identical forms of differential equations. Those forms become particularly evident if we drop the n-subscript and (**r**, t)-dependence notation. Thus, for the scalar $N_n(\mathbf{r}, t) \rightarrow N$, we have

$$\frac{1}{v}\frac{\partial N}{\partial t} = (\nu\,\Sigma_f - \Sigma_a)N + D\nabla^2 N \qquad (7.15b)$$

while for the scalar neutron flux $\phi_n(\mathbf{r}, t) \rightarrow \phi$, we similarly obtain

$$\frac{1}{v}\frac{\partial \phi}{\partial t} = (\nu\,\Sigma_f - \Sigma_a)\phi + D\nabla^2 \phi \qquad (7.15c)$$

It is therefore immaterial whether one seeks solutions for the neutron density or neutron flux since they satisfy identical forms of differential equations.

Equations (7.15) constitute the defining expression of interest for a finite non-reentrant medium and incorporate the important neutron gain/loss processes of fission, absorption and leakage. Once satisfactory initial and boundary conditions are specified, solutions can be sought and the reactor power determined according to Eq. (7.4).

7.3 Stationary Neutron Density

The determination of a solution of Eq. (7.15) of general applicability is not an elementary process and requires further consideration. For the special purpose of criticality assessment, we require $N_n(\mathbf{r}, t)$ to be stationary, that is, time independent. By this we mean that a steady-state neutron density is maintained which exactly balances the neutron production from fission against neutron losses by absorption and leakage. This implies

$$N_n(\mathbf{r}, t) \rightarrow N_n(\mathbf{r}) \,, \quad \therefore \; \frac{\partial}{\partial t} N_n(\mathbf{r}, t) = 0 \qquad (7.16)$$

for which we now seek the corresponding geometry and material composition. This steady-state imposition in Eq. (7.15) yields

$$\nu\,\Sigma_f\, v\, N_n(\mathbf{r}) - \Sigma_a\, v\, N_n(\mathbf{r}) + D\, v\, \nabla^2 N_n(\mathbf{r}) = 0 \qquad (7.17)$$

and may be written in more compact form as

$$\nabla^2 N_n(\mathbf{r}) + \left(\frac{\nu\Sigma_f - \Sigma_a}{D}\right) N_n(\mathbf{r}) = 0 , \tag{7.18a}$$

or, equivalently since the neutron speed v is a constant

$$\nabla^2 \phi_n(\mathbf{r}) + \left(\frac{\nu\Sigma_f - \Sigma_a}{D}\right) \phi_n(\mathbf{r}) = 0 . \tag{7.18b}$$

The formulation of a differential equation appropriate for a physical problem of interest is only part of an analysis; another part is the determination of a solution in explicit algebraic form which not only satisfies the mathematical requirements but also is in agreement with important physical features of the problem. Experience, insight and trial-and-error serve well in such work and frequently lead to voluminous results. In particular, Eq. (7.18) -- known as the homogeneous Helmholtz equation -- has been extensively studied with a considerable body of analysis existing in the mathematical physics and reactor physics literature. In order to provide a sufficiently compact solution analysis appropriate to our interest here, we will summarize the relevant mathematical and physical arguments in point form.

1. With ν, Σ_f, Σ_a and D as constants for the given reactor composition and the associated average neutron speeds, we write Eq. (7.18) as

$$\nabla^2 N_n(\mathbf{r}) + B^2 N_n(\mathbf{r}) = 0 , \tag{7.19a}$$

where

$$B^2 = \left(\frac{\nu\Sigma_f - \Sigma_a}{D}\right). \tag{7.19b}$$

This factor B^2 is frequently called the "buckling" of the reactor.

2. The solution of interest, $N_n(\mathbf{r})$, must be real and positive definite throughout the reactor. Since the defining equation, Eq. (7.18), is of 2nd order in **r**-space, two

boundary conditions or, alternatively, one boundary condition and a symmetry requirement on $N_n(\mathbf{r})$ must be imposed.

3. Common experience has shown that neutrons do escape from a bare reactor so that the neutron density everywhere on the surface $\mathbf{r}_b$ is positive, i.e. $N(\mathbf{r}_b)>0$. (Note that this requirement differs from that of many other "wave-like" problems when a zero boundary condition can be imposed.)

4. While the actual magnitude of $N_n(\mathbf{r}_b)$ is dependent upon arbitrary variables such as the reactor power, it has been found that a solution of Eq. (7.19a) extrapolated into the vacuum will attain a zero magnitude at a perpendicular distance $\mathbf{z}_0$ beyond the material boundary. Three features are associated with this "mathematical construct":

   4.1 the extrapolation distance $\mathbf{z}_0$ is dependent upon the material composition and neutron speed v but not upon reactor power;

   4.2 the solution $N_n(\mathbf{r})$ is very accurate throughout the medium except near the boundary;

   4.3 the boundary condition $N_n(\mathbf{r}_b+\mathbf{z}_0) = 0$ can be imposed on $N_n(\mathbf{r})$ of Eq. (7.19a).

Figure 7.4 has been designed to provide for the clarification of the above points.

7.4 The One-Dimensional Reactor

The points of the preceding discussions can be well illustrated and extended by an analysis of an idealized one-dimensional reactor. For this purpose, consider a homogeneous composition of fissile and other materials assembled into the shape of a slab of infinite extent in the y- and z-directions and of finite thickness in the x-directions. We position the x-axis so that $x=0$ is at the center of the slab, Fig. 7.5.

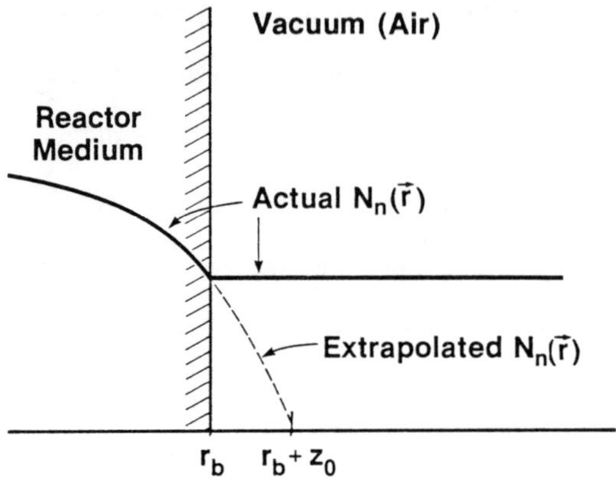

Fig. 7.4: Detail of the neutron density near a medium-vacuum boundary.

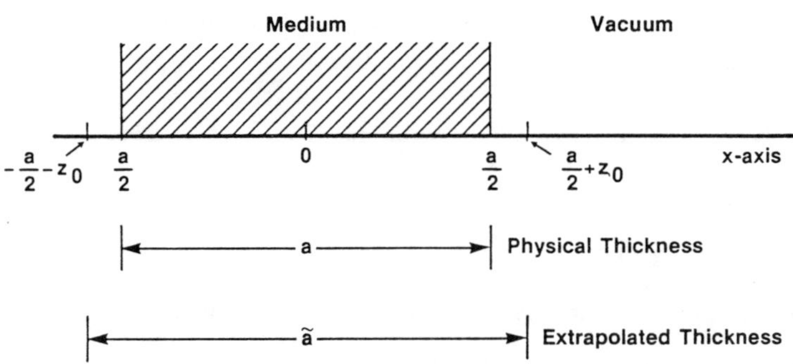

Fig. 7.5: One-dimensional slab reactor illustrating physical and extrapolated thicknesses.

With the reactor geometry thus defined, the neutron density $N_n(\mathbf{r})$ will not possess any y- and z-dependence; hence, we have $N_n(\mathbf{r}) = N_n(x)$ and therefore

$$\frac{d^2 N_n(x)}{dx^2} + B^2 N_n(x) = 0 . \tag{7.20}$$

We next consider appropriate boundary conditions for $N_n(x)$.

The incorporation of our extrapolation distance z_0 now makes the distinction between the physical thickness of the slab, represented by a, and the extrapolated thickness of the slab, given by $\tilde{a}$ and defined such that $\tilde{a}/2 = a/2 + z_0$, quite clear; the former is essential for physical reasons while the latter is useful for mathematical reasons. The boundary and symmetry conditions we now impose on the solution of Eq. (7.20) are

$$N_n(x) = N_n(-x) , \quad |x| < \frac{a}{2} , \tag{7.21a}$$

and

$$N_n\left(\frac{\tilde{a}}{2}\right) = 0. \tag{7.21b}$$

Further thought on Eq. (7.20) suggests that the functional form of the solution we want must have the property of repeating itself upon differentiation twice and concurrently change its sign. Well known functions with this property are

$$N_n(x) \sim \cos(Bx) \qquad \text{and} \qquad N_n(x) \sim \sin(Bx) , \tag{7.22}$$

We must dismiss the sin(Bx) function because it is not symmetric about $x=0$. A test solution which satisfies Eq. (7.20) is therefore, with A arbitrary,

$$N_n(x) = A\cos(Bx) , \tag{7.23}$$

Next, we impose the boundary condition Eq. (7.21b) on this proposed solution:

$$N_n\left(\frac{a}{2}\right) = A\cos\left(\frac{B\tilde{a}}{2}\right) = 0 . \tag{7.24}$$

Evidently, the argument of this cosine function must therefore assume any of the following infinite series of numbers:

$$\frac{B\tilde{a}}{2} = \left\{ \frac{\pi}{2}, \frac{3\pi}{2}, \frac{5\pi}{2}, \frac{7\pi}{2}, \ldots \right\} . \tag{7.25}$$

This is only possible if B is restricted to the following specific numerical values:

$$B_\ell = \frac{\ell\pi}{\tilde{a}}, \qquad \ell = 1, 3, 5, 7, \ldots . \tag{7.26}$$

The most general solution is therefore given by the summation of an infinity of linearly independent solutions

$$\begin{aligned} N_n(x) &= A_1 \cos\left(\frac{\pi x}{\tilde{a}}\right) + A_2 \cos\left(\frac{3\pi x}{\tilde{a}}\right) + A_3 \cos\left(\frac{5\pi x}{\tilde{a}}\right) + \ldots \\ &= \sum_\ell A_\ell \cos\left(\frac{\ell\pi x}{\tilde{a}}\right) \\ &= \sum_\ell A_\ell \cos(B_\ell x), \qquad \ell = 1, 3, 5, \ldots . \end{aligned} \tag{7.27}$$

The various terms $B_\ell = \ell\pi/a$ are called eigenvalues and the associated functions $\cos(B_\ell x)$ are known as eigenfunctions.

Two important questions now arise: first, what is the meaning of an infinite series for the neutron density $N_n(x)$, Eq. (7.27), when the reactor had been specified to operate at steady-state and, second, what is the meaning of B as a function of reactor thickness when, according to Eq. (7.19b), it should be a function of material composition.

The first question has long been resolved by actual measurements of $N_n(x)$ of experimental reactors operating at steady-state and found to contain only the leading term of Eq. (7.22),

$$N_n(x) = A_1 \cos\left(\frac{\pi x}{\tilde{a}}\right), \tag{7.28}$$

known as the fundamental mode, Fig. 7.6. Mathematical support is also provided for this result if a combined space-time dependent analysis is undertaken; it is then found that the higher order terms of Eq. (7.27) are associated only with transient conditions and quickly decay as the reactor approaches a steady state.

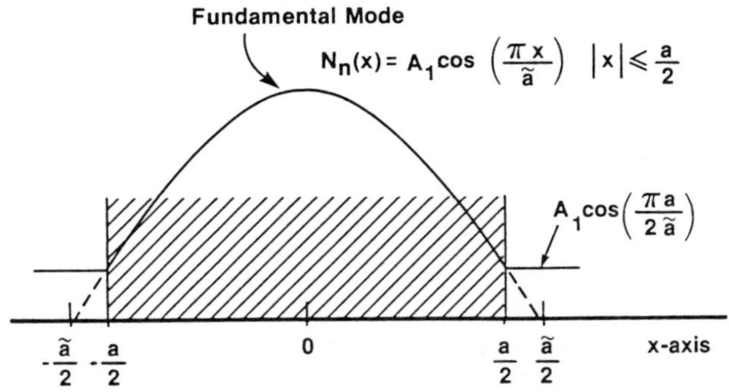

Fig. 7.6: Fundamental mode neutron density in a bare one-dimensional slab reactor.

We may therefore now compute the total recoverable fission power for our slab reactor per unit area, $\Delta y\ \Delta z$, according to Eq. (7.4) as follows:

$$P_{yz} = \sigma_f^f\, v\, Q^*_{fi}\, N_f \int_{-\frac{a}{2}}^{\frac{a}{2}} A_1 \cos\left(\frac{\pi x}{\tilde{a}}\right) dx$$

$$= 2\, A_1\, \frac{a}{\pi}\, \sigma_f^f\, v\, Q^*_{fi}\, N_f\, \sin\left(\frac{\pi a}{2\tilde{a}}\right) . \tag{7.29}$$

Typical power reactors possess dimensions on the scale of meters while extrapolation distances are of the order of several centimeters; thus $a >> z_0$ so that $\tilde{a} \sim a$ for most applications.

For reasons of completeness, we list the fundamental mode eigenfunctions for the above slab reactor as well as for the three general geometries of Fig. 7.1 in Table 7.1.

Table 7.1

Neutron density fundamental mode eigenfunctions and eigenvalues for the infinite slab and the three geometries of Fig. 7.1

Geometry	$N_n(r)$ Fundamental Eigenfunction	B^2 Fundamental Eigenvalue
Infinite slab	$\cos\left(\frac{\pi x}{\tilde{a}}\right)$	$\left(\frac{\pi}{\tilde{a}}\right)^2$
Spherical	$\frac{1}{r}\sin\left(\frac{\pi r}{\tilde{R}}\right)$	$\left(\frac{\pi}{\tilde{R}}\right)^2$
Rectangular	$\cos\left(\frac{\pi x}{\tilde{a}}\right)\cos\left(\frac{\pi y}{\tilde{b}}\right)\cos\left(\frac{\pi z}{\tilde{c}}\right)$	$\left(\frac{\pi}{\tilde{a}}\right)^2+\left(\frac{\pi}{\tilde{b}}\right)^2+\left(\frac{\pi}{\tilde{c}}\right)^2$
Cylindrical	$J_0\left(\frac{2.405\ldots r}{\tilde{R}}\right)\cos\left(\frac{\pi z}{\tilde{H}}\right)$	$\left(\frac{2.405\ldots}{\tilde{R}}\right)^2+\left(\frac{\pi}{\tilde{H}}\right)^2$

Note: (1) $J_0(\ )$ is the zeroth-order Bessel Function of the First Kind.

(2) The symbols x, y, z and r are coordinates whereas a, b, c, R and H are reactor dimensions.

7.5 Criticality Condition

We begin with an interpretation of the buckling B^2. Recall that the defining equation for the neutron density, Eq. (7.19), specified

$$B^2 = \frac{\nu\Sigma_f - \Sigma_a}{D}$$

$$= \frac{\nu\sigma_f^f N_f - \sum_i \sigma_a^i N_i}{D}, \qquad (7.30)$$

and thus defined the reactor buckling as a function of the important neutron-nucleus interactions and the nuclide composition of the reactor. Then, in our search for a solution of the defining equation, we found that at criticality, Eq. (7.26), the following must also hold for the one-dimensional slab reactor

$$B^2 = \left(\frac{\pi}{\tilde{a}}\right)^2 . \tag{7.31}$$

These two relationships provide the quantitative linkage between an intuitive recognition that geometry and composition must be rigidly related at criticality: that is, given a particular composition of fissile and other materials there can only be one slab thickness for which the neutron production is exactly balanced by neutron absorption and neutron leakage. The converse also holds. We conclude therefore that for the slab reactor, at criticality, we must have

$$\left(\frac{\pi}{\tilde{a}}\right)^2 = \frac{\nu \Sigma_f - \Sigma_a}{D} . \tag{7.32}$$

Consider next the following conceptual experiment. Supposing Eq. (7.32) holds and that somehow the slab thickness were reduced to a′ with no changes in the material composition. Hence, the right-hand part of Eq. (7.32) is unchanged and we get

$$\left(\frac{\pi}{\tilde{a}'}\right)^2 > \frac{\nu \Sigma_f - \Sigma_a}{D} . \tag{7.33}$$

However, the neutron leakage has not changed -- recall that neutron leakage is a surface phenomenon -- but the net effect of neutron multiplication and absorption, which are both volume phenomena, is to decrease the net number of neutrons which sustain the fission chain. Thus, the neutron density decreases with time and eventually decays to zero. This then defines a subcritical state.

Similarly, if the slab thickness were to be increased, the neutron population would tend to increased with time to define a supercritical state.

The tendency for fissile materials to become subcritical or supercritical can also be formulated in terms of a constant slab thickness in which the fissile concentration, leading to an increase in Σ_f, is conceived. The result is that the three cases of criticality interest can be defined using Eqs. (7.30) and (7.31) as follows:

$$\left(\frac{\pi}{\tilde{a}}\right)^2 > \frac{\nu\Sigma_f - \Sigma_a}{D} \quad : \quad \text{subcritical} \tag{7.34a}$$

$$\left(\frac{\pi}{\tilde{a}}\right)^2 = \frac{\nu\Sigma_f - \Sigma_a}{D} \quad : \quad \text{critical} \tag{7.34b}$$

$$\left(\frac{\pi}{\tilde{a}}\right)^2 < \frac{\nu\Sigma_f - \Sigma_a}{D} \quad : \quad \text{supercritical} \tag{7.34c}$$

By direct extension, we may replace these fundamental eigenvalues on the left of this tabulation by any of those listed in the last colum of Table 7.1 and arrive at criticality conditions for any of the corresponding geometries.

Note that a specific linkage between geometry and composition is of profound design importance. There may indeed be good reasons to specify either the material composition or the reactor core dimensions beforehand; the criticality condition then specifies the remaining parameters.

Discussion / Analysis 7.1

Examine the connection between components of a neutron (vector) current $\vec{J}(\vec{r})$ and the neutron (scalar) flux $\phi(\vec{r})$, both at $\vec{r}$.

For the graphics shown:

$$\phi(\vec{r}) = \int_\Omega F(\vec{r}, \vec{\Omega})\, d\vec{\Omega}$$

z, $\vec{r}$, x, y, $\vec{\Omega}$, a

distribution function of neutron "flow" through unit sphere about $\vec{r}$ and for various $\vec{\Omega}$ direction

1. number of neutrons crossing area a per s. $= N(\vec{r})v$
2. for isotropic "flow", $\phi(\vec{r})$ magnitude for all $\vec{\Omega}$ directions and hence $\phi/2$ cross in one direction and $\phi/2$ cross in other.

Take $\vec{\Omega}$ in $+z$-direction:

Now, the $\phi/2$ magnitude counts neutrons into and out of unit sphere. Hence, for isotropy

$$J_{+z}(\vec{r}) = \frac{1}{2}\left(\frac{\phi(\vec{r})}{2}\right) = \frac{\phi(\vec{r})}{4} = \text{partial neutron current in } +z \text{ direction}$$

Similarly: $J_{-z}(\vec{r}) = \dfrac{\phi(\vec{r})}{4}$

and $J_{+z}(\vec{r}) - J_{-z}(\vec{r}) = 0$

To think about:

Extend the above analysis to the case for which a flux gradient exists and show $J_{\pm z}(\vec{r}) = \dfrac{\phi(\vec{r})}{4} \mp \dfrac{D}{2}\dfrac{d\phi(\vec{r})}{dz}$

Discussion / Analysis 7.2

How should the radius and height of a cylindrical fission reactor be specified to minimize volume (and hence minimize fissile fuel requirements) ?

R

H

$$V = \pi R^2 H \quad , \quad R \simeq \tilde{R}, \; H = \tilde{H}$$

Criticality criteria (Sec. 7.5):

$$\underbrace{\left(\frac{2.405}{R}\right)^2 + \left(\frac{\pi}{H}\right)^2}_{\text{R \& H can vary (but equality must hold)}} = \underbrace{\frac{\nu \Sigma_f - \Sigma_a}{D}}_{\text{constant, say } B_m^2}$$

Solve for R^2:

$$R^2 = \frac{(2.405)^2}{B_m^2 - (\pi/H)^2}$$

∴ Volume = fc (H) : $V = \pi \left[\dfrac{2.405}{B_m^2 - (\pi/H)^2}\right] H$

Extremum H_o from $\left.\dfrac{dV}{dH}\right|_{H_o} = 0 \quad \Rightarrow \quad \boxed{H_o = \dfrac{\pi\sqrt{3}}{B_m}}$

Hence $R_o = \dfrac{2.405}{\left[B_m^2 - \left(\dfrac{\pi B_m}{\pi\sqrt{3}}\right)^2\right]^{1/2}} = \boxed{\dfrac{2.405}{B_m (2/3)^{1/2}}}$

Relation $R_o \simeq 0.55\, H_o$

To think about:

which of the three geometries of Fig 7.1 possesses the smallest critical mass ?

Problems

7.1 Confirm the solutions $N_n(\mathbf{r})$ and B^2, from Table 7.1, by substitution into Eq. (7.20). Remember to use the appropriate Laplacian for the relevant geometry.

7.2 Calculate the peak-to-average power ratio for the neutron densities for the cylindrical reactor Table 7.1.

7.3 Consider the following data for a hypothetical fissile assembly:

$$\nu = 2.2, \quad \Sigma_f = 0.010 \text{ cm}^{-1}$$

$$D = 1 \text{ cm}, \quad \Sigma_a = 0.021 \text{ cm}^{-1}$$

$$z_o = 0.6 \text{ cm}.$$

What is the critical radius for a bare spherical assembly possessing these parameters?

7.4 Examine some technical literature on the theory of particle transport in order to obtain a deeper understanding of a particle current, particle flux, and diffusion coefficient.

7.5 Why is B^2 called "the buckling"? Hint: Consider $-\nabla^2\phi/\phi$ and its associated radius of curvature in structural mechanics.

CHAPTER VIII

NUCLEAR ENERGY CONVERSION

Fission reactions in a fissile fuel domain constitute a source of energetic reaction products. The associated energy has to be extracted from the reactor and eventually transported to energy demand centers.

8.1 Energy Conversion

Primary energy sources seldom exist in the form and location where energy is required. Various forms of energy conversion and transport are therefore common and essential. As an example, consider energy from chemical sources. This form of energy is contained in a particular arrangement of hydro-carbon atoms largely located in natural underground reservoirs. This energy source has to be extracted, processed for different purposes, and finally transported to energy demand centers. There it is burned and the resultant energy, now in thermal form, used directly or further converted into forms such as linear motion -- as in vehicles -- or rotational motion, as in turbine-generators for energy in electrical form; this latter form involves a further transport of energy in an electrical grid.

Energy in electrical form has in recent decades been found to be particularly convenient and useful and most fission reactors serve the function of central station electrical power plants. This is accomplished by designing them as "exotic" boilers which convert nuclear energy into steam which then enters a steam-turbine to rotate an electric generator for electricity production. We depict this system concept in Fig. 8.1.

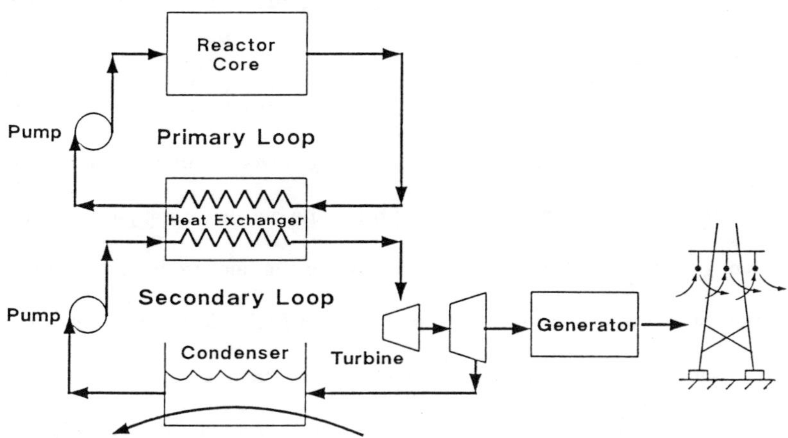

Fig. 8.1: Block and flow diagram of a fission reactor generating electrical energy.

When a fissile nucleus fissions, about 80% of the energy resulting from the mass decrement appears in the form of kinetic energy of the two massive fission products. As these high energy particles move in the fuel, they collide with numerous other atoms and eventually come to rest less than a millimeter from their origin. Along their short and irregular path of travel, the fission products generate considerable agitation and vibration among the host atoms which represents heat and hence raises the temperature of the fuel. As the local temperature increases, a heat current is established, leading to the transport of thermal energy from the fuel, through the cladding, and into the lower-temperature coolant. This coolant is constantly pumped through the core from where it exits at a higher temperature and enters a heat exchanger where its thermal energy is transferred to a secondary loop now carrying steam at high pressure. This steam then enters the high pressure end of a steam turbine in which the random thermal motion of the atoms and molecules, in their passage through the variously spaced turbine blades, is converted into rotational motion of the turbine shaft. Connected to this shaft is a system of conducting coils which then pass through magnetic fields. The resultant induced alternating current is then transmitted to an electrical distribution grid.

Some additional comments are appropriate. Figure 8.1 illustrates what is known as a "two-loop" system. The primary recirculating core-coolant is fully contained and totally separated from the secondary working fluid; the reason for not passing the primary coolant directly into the turbine, and thus bypassing the heat exchanger, is to minimize any radiation products which might escape from the fuel into the primary coolant and thereby eventually appear in the environment via the turbine or condenser. Note also, that rather than producing electricity, the secondary loop could supply steam for industrial processes or commercial and domestic space heating.

8.2 Reactor Thermodynamics

An examination of Fig. 8.1 suggests that the fission reactor core and associated components possess all the features of a thermodynamic system. For example, the reactor core is a "heat source", the turbine-generator is the "engine", the condenser is evidently a "heat sink", and the pump requires an "energy supply" to circulate the "working fluid" through the various system components. A thermodynamic analogue of Fig. 8.1 becomes particularly apparent if we assume the heat exchanger to be sufficiently efficient so that it can be ignored; we depict this in Fig. 8.2. On this figure we also display the control volume boundary and indicate the energy flow rates which cross this boundary.

With the reactor operating at steady-state and with no heat accumulation in any of the system components, we use the notation of Fig. 8.2 to write the basic power balance relationship as

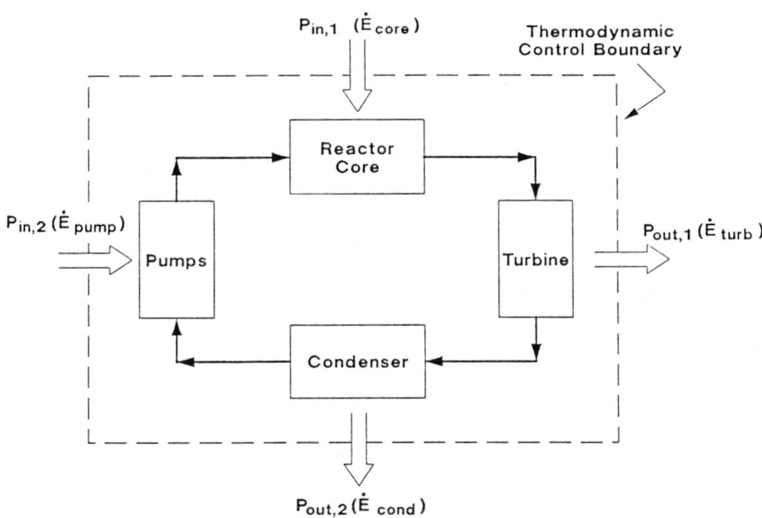

Fig. 8.2: Thermodynamic analogue of fission reactor system.

$$\dot{E}_{core} + \dot{E}_{pump} = \dot{E}_{turb} + \dot{E}_{cond} \, . \tag{8.1}$$

The generally most useful index of energy conversion efficiencies is given by the dimensionless parameter

$$\eta_{gross} = \frac{\dot{E}_{elect}}{\dot{E}_{core}} = \frac{\eta_{turb}\dot{E}_{turb}}{\dot{E}_{core}} \, , \tag{8.2}$$

where η_{turb} is the efficiency of the turbine-generator and $\dot{E}_{elect}$ is the electrical power delivered. Similarly,

$$\eta_{net} = \frac{\dot{E}_{elect} - \dot{E}_{pump} - \text{(ancillary power required)}}{\dot{E}_{core}} \, . \tag{8.3}$$

where ancillary power requirements involve various instrumentation and other devices of the plant. The electrical power delivered to a grid, $\dot{E}_{elect}$, is obtained by current and voltage measurements at the generator, while the nuclear power produced in the core, $\dot{E}_{core}$, is given by

$$\dot{E}_{core} = P_{fi}$$

$$= \gamma \int_{V_c} R_{fi}(r)\, Q^*_{fi}\, dV \ . \tag{8.4}$$

Here, $R_{fi}(r)$ is the fission rate at any point in the reactor and Q^*_{fi} is the thermally recoverable fraction of the fission Q-value and associated fission product decay heat; V_c is the fission core volume and γ is the conversion factor required to express MeV/s in units of watts (W).

Rather than computing $R_{fi}(r)$ throughout the core, it is more common to measure suitable average core inlet and outlet temperatures, T_{in} and T_{out}, of the coolant and, for the case of single-phase flow, determine P_{fi} from

$$P_{fi} = \dot{m}\, c_p(T_{out} - T_{in}) \ , \tag{8.5}$$

where $\dot{m}$ is the coolant mass flow rate and c_p is the mean specific heat capacity of the working fluid.

Thermal analyses of nuclear reactors are commonly undertaken using a thermodynamic function called the enthalpy represented by h. This function is defined at any point in a working fluid as an energy per unit mass of fluid by the relation

$$h = u + Pv \ . \tag{8.6}$$

Here u is the specific internal energy (J/kg), P is the local pressure (N/m^2) and v is the specific volume (m^3/kg). If h_{in} is the enthalpy for a unit mass of working fluid entering the core then the exiting enthalpy is

$$h_{out} = h_{in} + \int_{T_{in}}^{T_{out}} c_p(T)\, dT \ , \tag{8.7}$$

with $c_p(T)$ now the detailed temperature dependent specific heat capacity.

With the working fluid operating in a cycle, we may represent the pressure-enthalpy diagram as suggested in Fig. 8.3 below. The four energy conversion processes may be related to the enthalpy by extension of Eqs. (8.5) and (8.7). Specifically, we write as follows beginning with the pump:

$$\dot{E}_{pump} = \dot{m}\,(h_4 - h_1) \tag{8.8a}$$

$$\dot{E}_{core} = \dot{m}\,(h_2 - h_4) \tag{8.8b}$$

$$\dot{E}_{turb} = \dot{m}\,(h_2 - h_3) \tag{8.8c}$$

$$\dot{E}_{cond} = \dot{m}\,(h_3 - h_4) \tag{8.8d}$$

As previously used, $\dot{m}$ is again the coolant mass-flow rate.

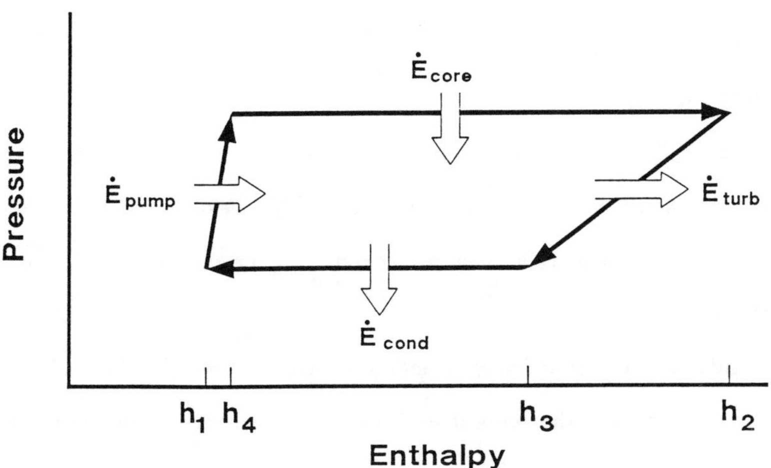

Fig. 8.3: Schematic of pressure-enthalpy diagram of the coolant working fluid.

In practice, the thermodynamic features of a fission reactor display considerable variations. For example, different working fluids such as gases and liquid metals may be used; at another level of design emphasis, numerous specialized design features such as superheat, reheat, tertiary loops, feed water heater, regenerative and others are used. All are introduced in order to increase the operational efficiency η_{gross}; typical values are in the range of 0.25 to 0.40.

We add that the core coolant inlet-outlet temperatures, T_{in} and T_{out}, are useful reference parameters. Thus, for an ideal system, we would have a Carnot efficiency of

$$\eta_o = \frac{T_{out} - T_{in}}{T_{out}} , \tag{8.9}$$

which is the maximum possible. The reason this ideal cannot be achieved is that actual reactors -- like all "real" thermodynamic systems -- do not operate reversibly between two isotherms and two adiabatics.

8.3 Fuel Heat Transfer

The preceding provided a general description of the nuclear-to-thermal-conversion and transport of energy of interest here. We next consider a more detailed discussion of some thermal relationships occurring in the primary energy production domain, namely the fuel.

Consider a typical cylindrical-geometry fuel cell consisting of the nuclear fuel, the surrounding cladding and the coolant designed to transport all the thermal energy generated in the fuel, Fig. 8.4. We may assume that the core is operating at steady state and that the cylindrical cell is radially uniform and of sufficient length so that end effects can be ignored. Our interest now is to establish the several analytical expressions which

---

Discussion / Analysis 8.1

In addition to P_{fi}, it is known that the decay of fission products contributes to reactor power. Examine these two components.

$$P_{tot}(t) = P_{fi}(t) + P_{dec}(t)$$

P_{fi}: $R_{fi}\, Q_{fi}^* = \sigma_f^f N_f N_n v Q_{fi}^*$

P_{dec}: $R_{dec}\, Q_{dec}^* \simeq \sum_i{}' \lambda_i N_i Q_i^*$ (i'th fission product)

Consider reactor scram at $t = 0$ (ie $k_\infty \to 0$)

$$P_{fi}(t) = P_{fi}(0)\, e^{-t/\tau_n^*} + P_{dec}(0) \sum_i{}' e^{-\lambda_i t}$$

Here $P_{tot}(0) = P_{fi}(0) + P_{dec}(0)$

and $P_{dec}(t) \sim 0.1\, P_{fi}(0)\, e^{-0.3t}$

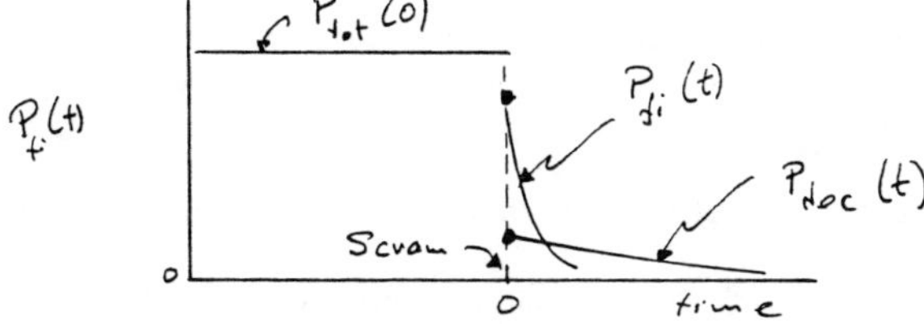

To think about:

What are the implications of decay power on continuing reactor cooling following a scram?

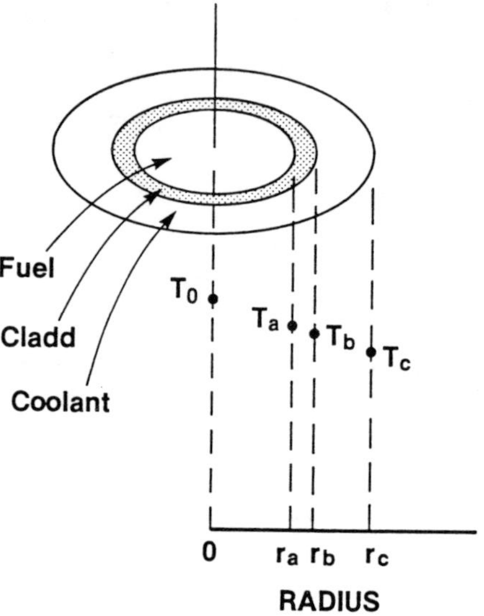

Fig. 8.4: Detail of a typical fission reactor fuel cell in cylindrical geometry.

relate nuclear energy production with such important properties as the temperature variations in the fuel cell and the heat current.

In Fig. 8.4, we indicate a typical unit volume of the fuel at an arbitrary radial distance r from the axis. Based on discussion of Chapt. 6, we can write the steady-state fission reaction rate as

$$\begin{aligned} R_{fi}(r) &= \sigma_f^f \, v \, N_f(r) \, N_n(r) \\ &= \Sigma_f^f(r) \, \phi_n(r) , \end{aligned} \tag{8.10}$$

where the symbols possess their usual meaning. The power density in this unit volume at radial coordinate r is evidently

$$\begin{aligned} P_{fi}(r) &= \sigma_f^f \, v \, N_f(r) \, N_n(r) \, Q_{fi}^* \\ &= \Sigma_f^f(r) \, \phi_n(r) \, Q_{fi}^* , \end{aligned} \tag{8.11}$$

with Q^*_{fi} used or defined in Eq. (8.4). Dimensionally, this power density $P_{fi}(r)$ is in units of MeV/m^3-s. In order to relate this fission-driven heat source to analyses more frequently found in the study of thermal energy systems, we introduce the power density in units of W/m^3;

$$q'''(r) = \gamma \, P_{fi}(r) \ . \tag{8.12}$$

Here, γ is again the ratio which converts MeV/s to watts (W), Eq. (8.4).

The symbol $q'''(r)$ is the fission power produced per unit volume at r and relates to its companion expression of fission power transported through a unit area, $\mathbf{q}''(r)$, by the general continuity equation

$$q'''(r) = \nabla \cdot \mathbf{q}''(r) , \tag{8.13}$$

where ∇ in the vector del-operator, here in cylindrical geometry. Also note that $\mathbf{q}''(r)$ is a vector quantity and represents the net heat current at radial coordinate r.

The reason for introducing the heat current $\mathbf{q}''(r)$ is that it now allows the introduction of an expression for the temperature at r using the Fourier law of heat conduction:

$$\mathbf{q}''(r) = -k_f \, \nabla T(r) \ . \tag{8.14}$$

Here, k_f is the local thermal conductivity in units of W/m^2-K. Substituting this equation in Eq. (8.13) and using Eq. (8.12) now provides for an analytical expression which relates the local fission power density to the local temperature:

$$\begin{aligned} \gamma \, P_f(r) &= \nabla \cdot \mathbf{q}''(r) \\ &= \nabla \cdot [-k_f \, \nabla T(r)] \ . \end{aligned} \tag{8.15}$$

Then, with k_f a constant, we write for cylindrical geometry with no axial or azimuthal dependence

$$\frac{1}{r} \frac{d}{dt} \left(r \frac{dT}{dr} \right) = -\left(\frac{\gamma}{k_f} \right) P_{fi}(r) \ . \tag{8.16}$$

This equation is most useful and particularly easy to solve for the fuel region, $0<r<r_a$, if the power density is constant throughout the fuel. Integrating twice, we get for the fuel temperature as a function of radial coordinate

$$T(r) = C_1 + C_2 \ln(r) - \left(\frac{\gamma P_{fi}}{k_f}\right)\frac{r^2}{4}\ , \tag{8.17}$$

where C_1 and C_2 are integration constants. Using $T(0) = T_o$ as the admissible maximum central-axis temperature in the fuel and also requiring symmetry about $r = 0$, yields immediately $C_1 = T_o$ and $C_2 = 0$, so that in the fuel region, a quadratic temperature profile exists:

$$T(r) = T_o - \left(\frac{\gamma P_{fi}}{4k_f}\right) r^2\ . \tag{8.18}$$

The use of the absolute value of the thermal energy current, $q''(r)$ of Eq. (8.14) and the use of Eq. (8.18) then leads to

$$q''(r) = \frac{1}{2}\,\gamma P_{fi}\ r\ . \tag{8.19}$$

Thus, the heat current at the central axis is zero and it then increases linearly with distance.

Equation (8.18) specifies that the temperature at the fuel-clad interface, T_a in Fig. 8.4, is given by

$$T_a = T_b - \left(\frac{\gamma P_{fi}}{k_f}\, r_a^2\right), \tag{8.20}$$

and can be used as a boundary condition to specify the temperature profile in the cladding. Some thought suggests that Eq. (8.16) also applies for the cladding domain $r_a<r<r_b$ providing one sets $P_{fi}(r) = 0$; that is, no fission energy is generated in the cladding. Formally, then, we write

$$\frac{1}{r}\frac{d}{dr}\left(r\,\frac{dT}{dr}\right) = 0\,, \qquad r_a \le r < r_b\,,\quad T(r_a) = T_a\ . \tag{8.21}$$

Solving this second order differential equation gives for the temperature in the cladding

$$T(r) = C_3 + C_4 \ln(r) \ , \tag{8.22}$$

with C_3 and C_4 integration constants. The informative result here is that the temperature in the cladding decreases logarithmically with r.

Finally, we consider the temperature variation from the outer-surface of the cladding into a distance of the coolant where it can be considered to be constant. At our level of interest, we accept Newton's law of cooling which asserts that at $r = r_c$, we have

$$q''(r_c) = h_{bc}[T_b - T_c] \ . \tag{8.23}$$

Here, $q''(r_c)$ is the magnitude of the heat current into the coolant, h_{bc} is the film coefficient of heat temperature, T_b is the outer-surface cladding temperature found from Eq. (8.22) and T_c is the average coolant temperature beyond the fluid film on the coolant. We depict the temperature profile and heat current for the cell of Fig. 8.4, as found in the above analysis, in Fig. 8.5.

Problems

8.1 Consult some reactor engineering texts and obtain specific pressure and temperature values for coolants of different reactor types.

8.2 Determine expressions for C_3 and C_4 in Eq. (8.22).

8.3 Derive the temperature profile for a homogeneous spherical fuel pebble.

8.4 Compare Eq. (8.14) with Eq. (7.14b) and describe their underlying physical similarities and differences.

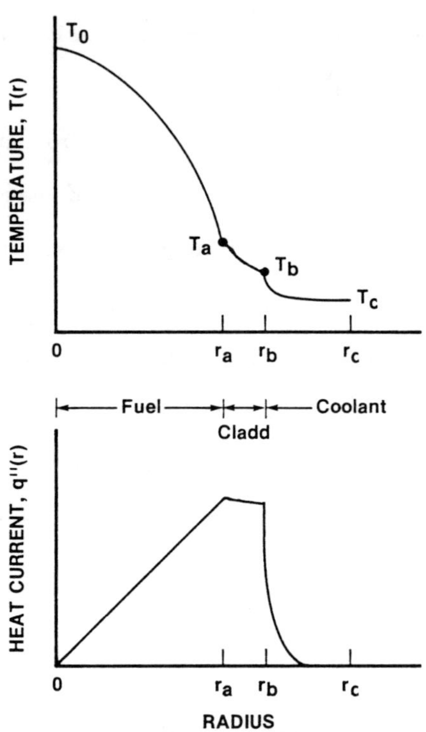

Fig. 8.5: Temperature and heat current profiles in a cylindrical fuel cell.

Discussion / Analysis 8.2

As a companion analysis to that of Sec. 8.3, outline how the axial temperature variation is determined.

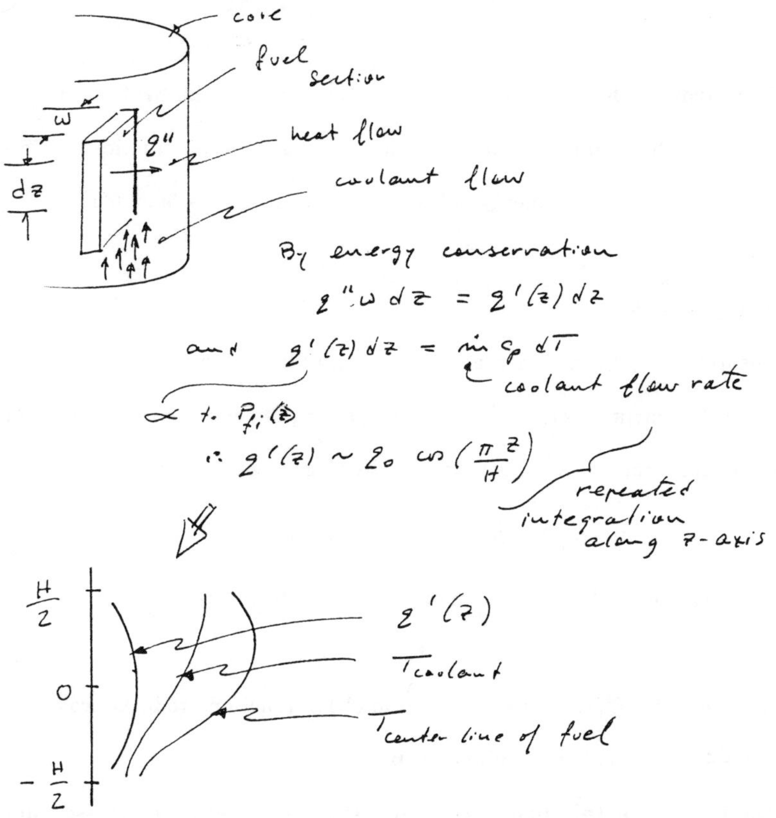

To think about

What temperature determines the limit of reactor thermal-to-electrical conversion efficiency?

CHAPTER IX

FUSION PHENOMENA

In contrast to the splitting of heavy nuclides, a significant release of energy also occurs when selected light nuclides interact to form heavier nuclides. This fusion approach to realizing nuclear energy holds much promise for the long term.

9.1 Fusion Reactions

Theoretical and experimental considerations have demonstrated that there exist numerous light element exoergic reactions. Of particular relevance to the first generation of fusion reactors is the process involving two hydrogen isotopes

$$^{2}_{1}H + ^{3}_{1}H \rightarrow (\ ) \rightarrow n + ^{4}_{2}He . \tag{9.1}$$

This equation is frequently written in more compact notation as

$$d + t \rightarrow n + \alpha , \tag{9.2}$$

where d is a deuteron ($^{2}{}_{1}H$) and t is the triton ($^{3}{}_{1}H$); reaction products are a neutron (n) and a helium-4 nucleus called the alpha particle (α).

The Q-value of this fusion reaction, that is the energy released due to the associated rest mass decrement, is calculated as follows:

$$\begin{aligned} Q_{fu} &= -(\Delta m)c^2 \\ &= [(m_d + m_t) - (m_n + m_\alpha)]c^2 \\ &= 17.6 \text{ MeV} . \end{aligned} \tag{9.3}$$

The next most accessible fusion reaction involves only deuterium and is characterized by two almost equally-likely reaction channels

$$d + d \rightarrow \begin{cases} p + t + 4.1\ \text{Mev} \\ h + n + 3.2\ \text{MeV} \ . \end{cases} \tag{9.4}$$

Here, p is a proton, and "h" stands for the helium-3 nucleus ($^3_2$He).

Deuterium is also known to fuse with other nuclei in both single and multiple channels:

$$d + h \rightarrow p + \alpha + 18.3\ \text{MeV} \tag{9.5a}$$

$$d + {}^6\text{Li} \rightarrow \begin{cases} p + \alpha + t + 2.6\ \text{Mev} \\ 2\alpha + 32.3\ \text{MeV} \\ \vdots \end{cases} \tag{9.5b}$$

The proton can also engage in exoergic fusion reactions:

$$p + {}^6\text{Li} \rightarrow h + \alpha + 4.0\ \text{MeV} \ , \tag{9.6a}$$

$$p + {}^{11}\text{B} \rightarrow 3\alpha + 8.7\ \text{MeV} \ . \tag{9.6b}$$

Numerous other fusion reactions are known to occur naturally and have also been studied under laboratory conditions.

9.2 Fusion Fuels

Considerable diversity of fusion reactions evidently exists. However, the existence of a fusion process is a necessary but not sufficient condition for its incorporation into an operating fusion energy system. Questions of the ready availability of the fuels, any complications in the sustainment of a sufficient interaction rate density, and the potential toxicity of the fusion fuels and reaction products need also to be considered.

The d+t reaction, Eq. (9.2), for example, leads to some important implications. Deuterium is a stable and plentiful fusion fuel existing in a ratio of 1 part in 6600 of the hydrogen nuclides in water; its supply may therefore be considered essentially "infinite" and the technology of extraction is well established. Tritium, on the other hand, is a radioactive beta emitter

$$t \rightarrow h + \beta^- , \tag{9.7}$$

with a half-life of 12.3 years, which is sufficiently short to be considered in the management of this nuclear fuel. Indeed, its natural supply is limited to very small quantities and is totally insufficient for a typical fusion reactor; hence, tritium has to be produced.

It is expected that the required tritium will be bred by neutron induced reactions in the blanket surrounding a fusion reactor. That is, the neutron produced in d+t fusion reaction, Eq. (9.2), would enter a lithium bearing domain and produce tritium by the following reactions:

$$n + {}^6\mathrm{Li} \rightarrow t + \alpha , \tag{9.8a}$$

$$n + {}^7\mathrm{Li} \rightarrow t + \alpha + n' . \tag{9.8b}$$

It is also proposed to extract tritium from the coolant of heavy water reactors where it is produced by neutron capture in deuterium,

$$n + d \rightarrow t . \tag{9.9}$$

Neutron-nucleus interactions thus also become relevant in the analysis of fusion reactors.

9.3 Thermonuclear Kinetics

In our discussion of fission reactor kinetics, it was found to be quite adequate to adopt a description in which one of the reacting particles -- the neutron -- migrates throughout a solid material composed of much more massive and stationary nuclei. The

Discussion / Analysis 9.1

How much energy would be released if all the ^2H (=d) in a liter of ordinary water could be made to fuse?

10 cm × 10 cm × 10 cm ⇝ $1\ \ell\ H_2O = 10^3\ g\ H_2O$

with $18\ g\ H_2O \Rightarrow 0.6 \times 10^{24}\ H_2O$ molecules

$10^3\ g\ H_2O \Rightarrow 3.3 \times 10^{25}\ H_2O$ molecules

↳ 6.6×10^{25} hydrogen atoms

@ 0.015 % deuterium abundance in hydrogen

↳ $N_d \approx 10^{22}$ deuterons in $1\ \ell$ of H_2O

From Eq. (9.4): $\bar{E}_d = \frac{4.1 + 3.2}{4} = 1.8$ MeV/deuteron.

Total d+d fusion energy

$$E_{tot} = 1.8 \left(\frac{MeV}{deuteron}\right) \cdot 10^{22} \left(\frac{deuterons}{liter}\right)$$

$$= 1.8 \times 10^{22}\ MeV/\ell = \boxed{2.9 \times 10^9\ J/\ell}$$

Compare: $1\ \ell$ gasoline $\sim 4 \times 10^7\ J/\ell$

To think about:
How big a lake is required so that the small fraction of deuterium in it would yield as much energy as estimated for all global oil resources?

underlying physical basis for this description is that the neutron is uncharged and hence unaffected by electrostatic forces from the atomic electrons or from the protons in the nucleus; it is thus free to wander about until it happens to be absorbed by a nucleus possessing a cross section for this neutron removal process.

The physical basis for fusion reactions is quite different. First, the interacting particles are of comparable mass. Then, at nominal temperatures, the orbiting electrons prevent close contact between the nuclei. Raising the medium temperature to increase the kinetic energy of the atoms will ionize the hydrogen atoms -- only 13.6 eV is required to remove an electron. When most of the atoms have become ionized, the medium will consist only of charged nuclides and charged electrons thus forming a plasma, sometimes called the Fourth State of Matter. In this state, it is possible for the different types of ions to be represented as being in equilibrium at some elevated temperature.

The dominant forces in effect in a domain containing ions and electrons are electrostatic. Coulomb repulsion now governs the likelihood of deuterium fusion with tritium. At a distance of separation r, the force of repulsion between hydrogen ions is given by

$$F = \frac{1}{4\pi\varepsilon_o} \frac{q^2}{r^2} . \tag{9.10}$$

where ε_0 is the permittivity of free space and q is the electronic charge. The total energy required to bring the two hydrogen ions into sufficiently close contact for the nuclear forces of attraction to become dominant is the Coulomb barrier given by

$$E_B = \frac{1}{4\pi\varepsilon_o} \frac{q^2}{(r_d + r_t)} . \tag{9.11}$$

For d+t fusion this is of the order of 300 keV; tunneling, however, does allow for some fusion to occur even at lower energies.

The above suggests that, in order to bring about fusion in hydrogen, one must first ionize a suitable deuterium-tritium mixture of gas and then raise the energy of the ions. Various mechanisms may be employed for the latter purpose: electromagnetic waves, induction of a current flow by an external magnetic field, injection of particles from an accelerator, and others.

The consequence of heating is the establishment of a range of kinetic energies for the ions and electrons. This energy spectrum will, under equilibrium conditions, define a distribution known as the Maxwell-Boltzmann distribution and given by

$$M(E) = \frac{2}{\sqrt{\pi}} \frac{1}{kT} \left(\frac{E}{kT} \right)^{1/2} \exp\left(-\frac{E}{kT} \right) . \tag{9.12}$$

Here, k is the Boltzmann constant and T is the absolute temperature (in units of kelvin) for the appropriate charged particle specie; the normalization factor here is specified according to

$$\int_0^\infty M(E)\, dE = 1 . \tag{9.13}$$

Figure 9.1 provides a graphic depiction for M(E) for two temperatures.

The availability of a distribution function such as Eq. (9.12) can aid considerably in the energetics analysis of a fusing medium. For example, the average energy of the ions is found to be

$$\begin{aligned} \overline{E} &= \int_0^\infty E\, M(E)\, dE \\ &= \frac{3}{2}\, kT . \end{aligned} \tag{9.14}$$

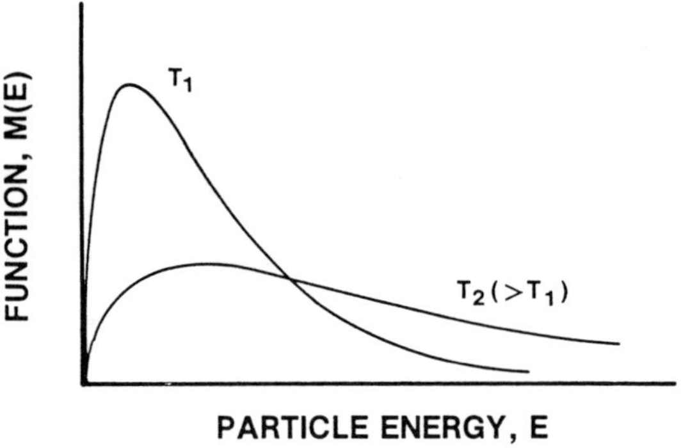

Fig. 9.1: Particle distribution as a function of energy for two different temperatures.

That is, the average energy of the ions is directly proportional to the temperature associated with the energy distribution of the ensemble of particles. We conclude, therefore, that to increase the energy of N particles to a specified temperature T -- assuming they possess little initial kinetic energy and supposing that little energy is lost during the process of heating -- requires an energy input E_{in} given by

$$\begin{aligned} E_{in} &= N\overline{E} \\ &= \frac{3}{2}\, NkT \, . \end{aligned} \tag{9.15}$$

This close connection between a particular energy state and the temperature as a characteristic for a distribution of ion energies has led to their use as an interchangeable parameter. Referring to Eq. (9.12), it is evident that the exponent represents a dimensionless parameter in which energy is measured in units of kT. This equivalence, then, for a constant factor k can be used to equate

$$1 \text{ eV} \Leftrightarrow 11{,}609 \text{ K} \ , \tag{9.16}$$

and any of its multiple equivalents. It is in this sense that a 10 keV plasma is at the very high temperature of 1.16×10^8 K.

9.4 Sigma-V Parameter

Expressions for the fusion reaction rate involving two ions can be established by extension of the neutron-nucleus interaction analysis of Chapt. 5. Even though Coulomb forces now enter into the detailed mechanism, a similar approach may nevertheless be taken.

Consider two intersecting beams of d and t ions, Fig. 9.2. The deuterons in one beam are characterized by a particle density N_d and all possess the same speed and direction $\mathbf{v}_d$; a corresponding description applies to the tritium beam, N_t and $\mathbf{v}_t$. Fusion reactions occur in the overlapping region and are detected by the appearance of the fusion products n and α, Eq. (9.2).

Heuristic considerations suggest that the fusion reaction rate, R_{fu}, will be proportional to the deuterium ion density N_d, to the tritium ion density N_t, and to the relative speed of these two beams v_r. Hence, we write this proportionality as

$$R_{fu} \propto v_r N_d N_t \ , \tag{9.17a}$$

where

$$v_r = |\mathbf{v}_d - \mathbf{v}_t| \ . \tag{9.17b}$$

This proportionality relationship can be converted into an equation by the introduction of a proportionality constant specifically applicable to the particles involved and their relative speed:

$$R_{fu} = \sigma_{dt}(v_r) \, v_r N_d N_t \ . \tag{9.18}$$

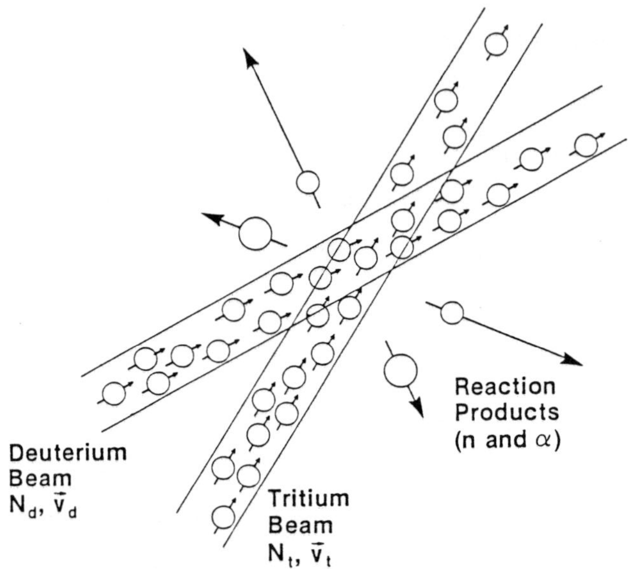

Fig. 9.2: Intersection of a deuterium and a tritium beam with fusion reactions d + t occurring in the overlapping region.

Here, $\sigma_{dt}(v_r)$ is the microscopic fusion cross section and possesses units of area and is a function of the scalar v_r. (Compare this analysis to that of neutron-nucleus interaction of Chapt. 5.)

Equation (9.18) is highly specialized since it applies to intersecting monoenergetic beams which are generally not encountered in fusion energy systems. As suggested in the preceding section, the ions will possess a range of energies and hence a range of speeds; also, their direction of motion will largely be random. What is needed, therefore, is an equation of the form of Eq. (9.18) but which integrates all possible ion directions of motion and all possible speeds of the ions. The integral calculus is well-suited for this extension.

To generalize Eq. (9.18), we first take the particle densities N_d and N_t to possess a range of speeds and directions:

$$N_d \rightarrow N_d f(\mathbf{v}_d) \tag{9.19a}$$

$$N_t \rightarrow N_t g(\mathbf{v}_t) \ . \tag{9.19b}$$

Here, $f(\mathbf{v}_d)$ and $g(\mathbf{v}_t)$ are normalized distribution functions in velocity space with N_d and N_t the total deuterium and tritium densities. Then, with the ions as function of velocity and with v_r and $\sigma_{dt}(v_r)$ also functions of $\mathbf{v}_d$ and $\mathbf{v}_t$, Eq. (9.18), we integrate over all ranges of these two independent vectors to obtain the total fusion reaction in a unit volume:

$$\begin{aligned} R_{fu} &= \int_{\mathbf{v}_d}\int_{\mathbf{v}_t} \sigma_{dt}(v_r)\, v_r\, N_d f(\mathbf{v}_d)\, N_t g(\mathbf{v}_t)\, d\mathbf{v}_d\, d\mathbf{v}_t \\ &= N_d N_t \int_{\mathbf{v}_d}\int_{\mathbf{v}_t} \sigma_{dt}(v_r)\, v_r\, f(\mathbf{v}_d)\, g(\mathbf{v}_t)\, d\mathbf{v}_d\, d\mathbf{v}_t \ . \end{aligned} \tag{9.20}$$

The integral could be evaluated for known $\sigma_{dt}(v_r)$, $f(\mathbf{v}_d)$, and $g(\mathbf{v}_d)$ appropriate for the case of interest; indeed, if the ion densities possess a Maxwell-Boltzmann energy distribution, Eq. (2.12), then it can be derived from M(E) with the aid of the appropriate Jacobian of the transformation. In any case, the integral possesses the form of a "weighted mean" for the product $\sigma_{dt}(v_r)v_r$ and, hence, we define

$$<\sigma v>_{dt} = \int_{\mathbf{v}_d}\int_{\mathbf{v}_t} \sigma_{dt}(v_r)\, v_r\, f(\mathbf{v}_d)\, g(\mathbf{v}_t)\, d\mathbf{v}_d\, d\mathbf{v}_t \ , \tag{9.21}$$

and call it the "sigma-vee" reaction parameter.

We recall, however, that M(E) is a function of the medium temperature and therefore must expect that $<\sigma v>_{dt}$ possesses such a dependence. We depict this dependency in Fig. 9.3 and provide a detailed tabulation in Appendix A, Table A.6.

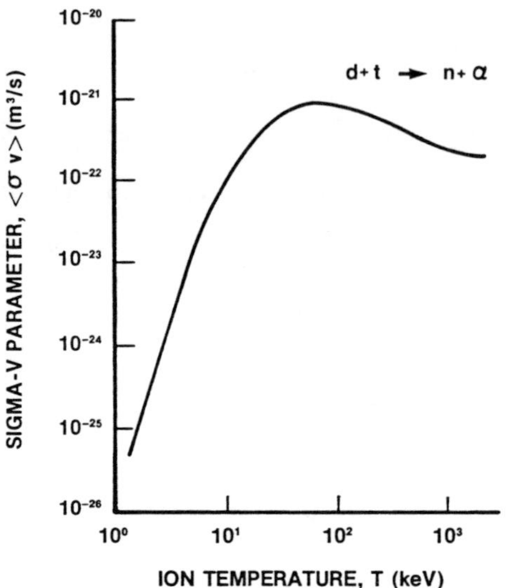

Fig. 9.3: Maxwellian averaged sigma-v reaction parameter for d + t fusion.

The fusion reaction rate for deuterium-tritium ions at a given temperature may thus be compactly written as

$$R_{fu} = <\sigma v>_{dt} N_d N_t \ , \tag{9.22}$$

and the rate at which fusion energy is released is

$$P_{fu} = <\sigma v>_{dt} N_d N_t Q_{dt} \ , \tag{9.23}$$

where Q_{dt} is the Q-value for the d + t fusion reaction, 17.6 MeV.

Note the essential similarity between the descriptions of ion-ion (fusion) reactions and neutron-nucleus (fission) reactions, Eq. (9.22) and Eq. (5.13): the reaction rate density involves the product of the two interacting particle densities, their relative speeds and the microscopic cross sections for the specific type of reaction of interest. The reason for the differences in algebraic form is attributable to historical preferences of accommodating kinematic differences for cases of interest.

Problems

9.1 Confirm the correctness of Eq. (9.14). (Hint: Integrate by parts.)

9.2 Determine the most frequently occurring particle energy for a Maxwellian distribution as a function of temperature.

9.3 Consider a 50-50% mixture of deuterium and tritium with a total ion density of 10^{14} cm^{-3}. Calculate the fusion power density for the temperature range from 10 keV to 100 keV. Express P in units of MW/m^3.

9.4 Given the reaction

$$p + {}^9\mathrm{Be} \rightarrow \begin{cases} \xrightarrow{.8} \alpha + {}^6\mathrm{Li} + 2.1\ \mathrm{MeV} \\ \xrightarrow{.2} 2\alpha + d + 0.8\ \mathrm{MeV} \end{cases}$$

and an initial inventory of 10^7 beryllium-9 nuclei, how many He-4 could be produced and what is the total energy released?

Discussion / Analysis 9.2

How is the Q_{dt} energy in the reaction $d + t \rightarrow n + \alpha$ shared among the reaction products?

before fusion — after fusion

(d) → · ← (t) ← (n) · (α) →

For negligible initial $E_{d,k}$ and $E_{t,k}$:

Energy balance: $\frac{1}{2} m_n v_n^2 + \frac{1}{2} m_\alpha v_\alpha^2 = Q_{dt}$ (1)

Momentum balance: $m_n v_n = m_\alpha v_\alpha$ (2)

For $E_{k,n}$, eliminate v_α in (1):

$$\therefore \frac{1}{2} m_n v_n^2 = \frac{Q_{dt}}{\left(1 + \frac{m_n}{m_\alpha}\right)} \simeq \frac{17.6}{1 + \frac{1}{4}} = \boxed{14.1 \text{ MeV}}$$

For $E_{k,\alpha}$, eliminate v_n in (1):

$$\therefore \frac{1}{2} m_\alpha v_\alpha^2 = \frac{Q_{dt}}{\left(1 + \frac{m_\alpha}{m_n}\right)} \simeq \frac{17.6}{1 + \frac{4}{1}} = \boxed{3.5 \text{ MeV}}$$

ie a 80% - 20% energy assignment

To think about:

For $p + {}^{11}B \rightarrow 3\alpha$, how is Q_{pB} shared among the three alphas?

CHAPTER X

FUSION ENERGETICS

The production and utilization of energy from fusion processes require special devices and careful accounting of energy. We discuss some of these and relate them to the overall energy viability of fusion reactor systems.

10.1 Fusion Devices

The simplest fusion device is a light-ion accelerator with a solid target of appropriate light elements. Detectors suitably placed around the target would then provide evidence of fusion events by recording the appropriate emitted reaction products and their energies. While fusion may indeed be readily demonstrated in such a device, it has long been known that the dominance of ionization effects and of Rutherford scattering -- that is, deflection of the ions by electrostatic repulsion -- is sufficiently pronounced that most ions slow down to below their Coulomb barrier, Eq. (8.11); hence, relatively few fusion reactions occur and such a systems concept can be shown not to be energetically viable.

The most widely researched approach leading to the design of a fusion reactor is based on the confinement of ions by magnetic field effects. In this approach, one begins with a hydrogen gas containing deuterium and tritium. Energy is then supplied to ionize the gas and also heat the ions to a sufficiently high temperature -- typically in the 10^7 K range. Then, in order to provide repeated opportunity for the high temperature ions to collide and fuse, the ions are confined by magnetic fields. This magnetic confinement approach is based on the known phenomenon that ions tend to spiral about magnetic field lines as described by the Lorentz force equation. By a suitable shaping of the magnets which produce the magnetic field lines, one seeks to contain the ions in a specified region at a desired ion density. For example, a current-carrying solenoid creates axial field lines which will cause an ion to move in a spiral pattern with their thermal energy and scattering collisions providing for some axial motion. Then, to reduce ion escape through the ends, the magnetic field is increased with the result that the magnetic field curvature induces an axial component of deceleration and eventual partial reflection of the ion back toward the central plasma region. Such devices are called magnetic mirrors, Fig. 10.1, and hence the attainment of fusion by such methods is called magnetic confinement fusion (MCF).

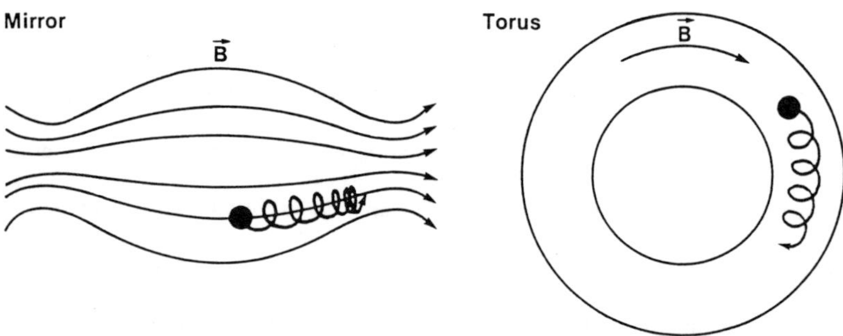

Fig. 10.1: Schematic depiction of a magnetic mirror and a magnetic torus. The pattern of magnetic-field lines and ion trajectories is also suggested.

Since no magnetic mirror can be totally effective in preventing end-leakage of high energy ions, the concept of shaping a solenoid into a "doughnut" form was introduced, Fig. 10.1. This avoids end losses but, because of the consequent radial variations of the magnetic field in the torus, other ion confinement problems arise. These adverse effects may be minimized by the superposition of suitable poloidal fields with toroidal magnetic fields. The label "Tokamak" (a word of Russian origin and resulting from a contraction of toroidal-chamber-magnetic) is associated with such devices if ion heating occurs by a current flow in the plasma induced by transformer action.

While the above approaches are based on the confinement of ions by magnetic forces, another approach is based on inertial forces in order to attain the same objective. In this approach, one begins with a suitably layered pellet of about 5 mm radial dimension and containing the fusion fuel at its center. An intense pulse of radiation -- such as a laser beam or high energy ions -- strikes the pellet from several evenly spaced directions. The initial energy deposition in the outer layer causes the material to evaporate and spread out in all directions with the result that an outward directed ablation process and a consequent inward directed pressure wave occurs. This centrally directed pressure wave causes heating, ionization, and a significant increase in particle density at the center causing fusion to occur followed by pellet disassembly. This approach is called inertial confinement fusion (ICF) and is graphically depicted in Fig. 10.2 below.

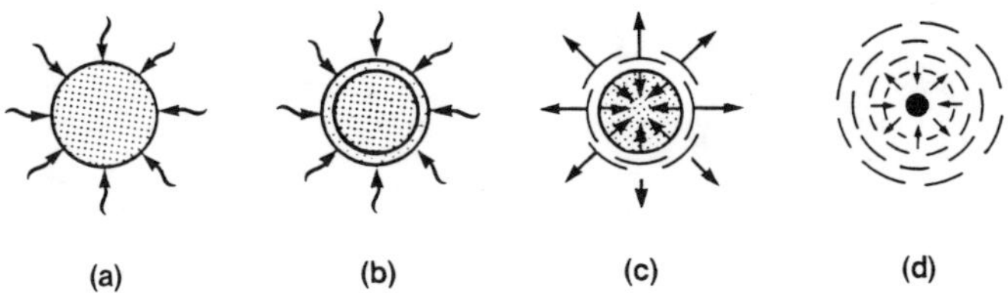

Fig. 10.2: Illustration showing the sequential processes associated with inertial confinement fusion.

Discussion / Analysis 10.1

Analysis of motion of an isolated ion in a uniform magnetic field $\vec{B}$

The dynamics is governed by the Lorentz Force Eq.

$$m \frac{d\vec{v}}{dt} = q(\vec{v} \times \vec{B})$$

mass of ion (m); charge of ion (q); velocity of ion ($\vec{v}$)

Orient the (x, y, z) coordinate system such that $\vec{B}$ is parallel to z-axis.

z, y, x axes; $\vec{B}$ along z

Here, $B_x = B_y = 0$ and Lorentz Eq. now has components:

$$\frac{dv_x}{dt} = \left(\frac{qB_z}{m}\right) v_y \; ; \quad v_x(0) = v_{x,0} \qquad (1)$$

$$\frac{dv_y}{dt} = -\left(\frac{qB_z}{m}\right) v_x \; ; \quad v_y(0) = v_{y,0} \qquad (2)$$

$$\frac{dv_z}{dt} = 0 \; ; \quad v_z(0) = v_{z,0} \qquad (3)$$

easy part: $v_z(t) = v_{z,0}$
(by integration)

$$\therefore z(t) = z_0 + v_{z,0}\, t$$
$$= z_0 + v_{\parallel} t$$

Define gyrofrequency $\omega = \left(\frac{qB_z}{m}\right)$, a constant

cont'd

Re $v_x(t)$ and $v_y(t)$: eliminate coupling in Eqs. (1) & (2) by differentiation and substitution to get

$$\frac{d^2 v_x}{dt^2} + \omega^2 v_x = 0 \quad ; \quad \frac{d^2 v_y}{dt^2} + \omega^2 v_y = 0$$

Solutions : $v_i(t) = a_i \cos(\omega t) + b_i \sin(\omega t)$

Impose : diamagnetic property of plasma

$$a_x = -b_y \equiv v_\perp \quad ; \quad a_y = b_x = 0$$

$$v_x(t) = v_\perp \cos(\omega t)$$

$$\hookrightarrow x(t) = x_0 + \frac{v_\perp}{\omega} \sin(\omega t)$$

$$v_y(t) = -v_\perp \sin(\omega t) \qquad \Bigg\} \; r = \frac{v_\perp}{\omega}$$

$$\hookrightarrow y(t) = y_0 - \frac{v_\perp}{\omega} \cos(\omega t)$$

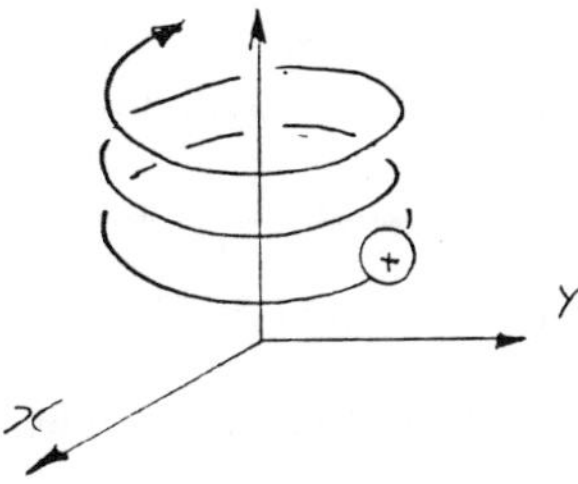

"spiral" about $\vec{B}$ lines

To think about :

Prove that the "pitch of the spiral" is a constant.

10.2 Power/Energy Components

Both energy and power constitute quantities which enter or leave a domain of interest. For example, energy may be injected into a fusion reactor core domain and result in increased thermal motion of the particles; the rate at which this energy is injected is the power. Various such transfers of energy and power are associated with a fusing medium and need to be included in the energy analysis of fusion energy systems.

Consider a density N_d and N_t for d + t fusion. These ions will fuse at the rate of

$$R_{fu} = <\sigma v>_{dt} N_d N_t \ , \tag{10.1}$$

where $<\sigma v>_{dt}$ is the sigma-vee parameter of Sect. 9.4. Each such reaction releases Q_{dt} units of energy due to the associated mass decrement. Hence, the fusion power, that is the rate at which fusion energy is released, is

$$\begin{aligned} P_{fu} &= \frac{dE_{fu}}{dt} \\ &= R_{dt} Q_{dt} \\ &= <\sigma v>_{dt} N_d N_t Q_{dt} \ . \end{aligned} \tag{10.2}$$

Additionally, it is known that ions at high temperature will be buffeted about because of the numerous scattering effects from other ions and electrons. This leads to ion and electron deceleration and the emission of bremsstrahlung radiation which represents the transformation of energy from kinetic to electromagnetic form. A suitable expression for this power component is given by

$$P_{br} \simeq A[N_d + N_t]N_e(kT_e)^{1/2} \ , \tag{10.3}$$

where A is a constant, N_e is the electron density and T_e is the electron temperature.

Another important power/energy component is specifically associated with the thermal motion of the ions and electrons. If a Maxwell-Boltzmann distribution applies, then the average thermal energy of the j-type ions, Sec. 9.3, is found to be

$$\bar{E}_j = \frac{3}{2} k T_j \ , \tag{10.4}$$

where T_j refers to the particular temperature of the j-type ions. The total thermal energy involving N_d and N_t ions, and N_e electrons is therefore

$$E_{th} = \left(\frac{3}{2}\,k\,T_d\right)N_d + \left(\frac{3}{2}\,k\,T_t\right)N_t + \left(\frac{3}{2}\,k\,T_e\right)N_e\,. \tag{10.5}$$

This thermal energy state will have been attained upon the injection of power over a period of, say τ units of time. We associate a power requirement with this energy component of

$$E_{th} = \int_0^{\tau} P_{th}\,dt\,, \tag{10.6}$$

and if the power supplied is a constant over the period τ, then

$$E_{th} = P_{th}\,\tau\ . \tag{10.7}$$

These power and/or energy expressions will be used in our further energetics analysis of a fusing system.

10.3 Lawson Criterion

The Lawson Criterion is a compact statement about the energy viability of a fusing system in terms of some basic parameters. It rests on the simple premise that, over a suitable cyclic interval, the total energy supplied to a fusion system must be less than the total energy extracted. That is,

$$E_{out} > E_{in}\ , \tag{10.8}$$

so that $E_{net} = E_{out} - E_{in} > 0$. Our interest now is to obtain various components for E_{in} and E_{out} in terms of the analysis of the preceding section. An important feature is that energy transformations never occur with 100% efficiency so that the conversion efficiencies, η_i for a i-process, needs to be included. A typical time cycle of τ units of time will be implied for the inequality, Eq. (10.8).

Consider Fig. 10.3 to represent a magnetic confinement fusion system operating in a pulsed mode. Here we suggest that an amount of energy E_{in} is initially supplied to provide the necessary temperature conditions. The fraction $\eta_{in} E_{in}$ appears in the form of bremsstrahlung energy, E_{br}, and thermal energy, E_{th}; we therefore write

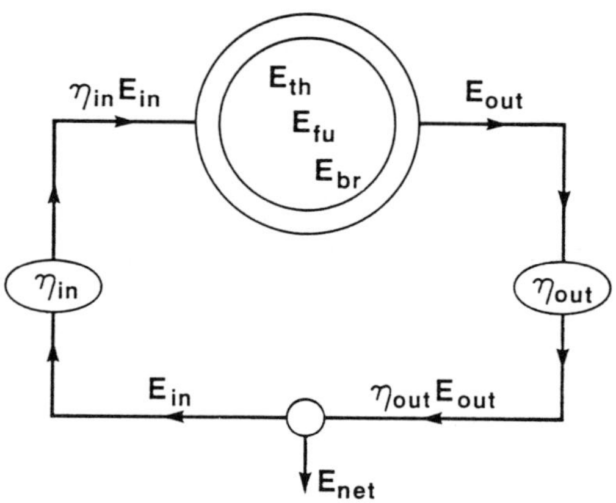

Fig. 10.3: General schematic of energy flow in a fusion energy system.

$$\eta_{in} E_{in} = E_{th} + E_{br} \; . \tag{10.9}$$

The energy potentially extractable will only be a fraction of the thermal energy, $\eta_{th} E_{th}$, a fraction of the bremsstrahlung energy, $\eta_{br} E_{br}$, and a fraction of the energy resulting from the fusion reactions, $\eta_{fu} E_{fu}$; that is

$$E_{out} = \eta_{th} E_{th} + \eta_{br} E_{br} + \eta_{fu} E_{fu} \; . \tag{10.10}$$

The essential inequality, Eq. (10.8), together with Eqs. (10.9) and (10.10), now becomes specifically

$$E_{th} + E_{br} < \eta_{in} \eta_{out} [E_{th} + E_{br} + E_{fu}] \; , \tag{10.11}$$

where, for reasons of algebraic convenience, we have taken

$$\eta_{out} = \eta_{th} = \eta_{br} = \eta_{fu} \cdot \qquad (10.12)$$

We take τ as the length of a typical operating cycle and assume that the power flows can be considered as constants during this interval. Introducing the particle densities and their temperature into each of the above expressions according to the discussion of the preceding section gives the following:

$$1.:\ E_{th} = \left(\frac{3}{2}k\right)[T_d N_d + T_t N_t + T_e N_e], \qquad (10.13a)$$

$$2.:\ E_{br} = P_{br}\tau$$

$$= A[N_d + N_t]N_e(kT_e)^{1/2}\tau, \qquad (10.13b)$$

$$3.:\ E_{fu} = P_{fu}\tau$$

$$= \langle\sigma v\rangle_{dt} N_d N_t Q_{dt}\tau, \qquad (10.13c)$$

We further propose that the fusing media consist of equal densities of deuterium and tritium so that an ion density N can be introduced which relates to the other ion densities by

$$N = 2N_d = 2N_t = N_e \ . \qquad (10.14a)$$

Finally, a thermal equilibrium state for all particles, gives further

$$T = T_d = T_t = T_e \ . \qquad (10.14ab)$$

Equation (10.11) now becomes

$$\left\{3kTN + AN^2(kT)^{1/2}\tau\right\} < \eta_{in}\,\eta_{out}\left\{3kTN + AN^2(kT)^{1/2}\tau + \frac{\langle\sigma v\rangle_{dt}}{4}N^2Q_{dt}\tau\right\} \qquad (10.15)$$

Cancelling N in each term and solving for $N\tau$ yields

$$N\tau > \frac{3kT(1 - \eta_{in}\,\eta_{out})}{\eta_{in}\,\eta_{out}\left[A(kT)^{1/2} + \frac{\langle\sigma v\rangle_{dt}}{4}Q_{dt}\right] - A(kT)^{1/2}} \qquad (10.16)$$

Every term on the right-hand side is either a constant or a function of temperature of the fusing media of interest. This relation has been evaluated for d + t and similarly for d + d fusion as a function of kT and is displayed in Fig. 10.4. Thus, if τ is small, then N must be large; conversely, if τ is larger, N can be small. In any case, their product must satisfy Eq. (10.16) and, as suggested in Fig. 10.3, this is easier for d + t fusion at kT $\simeq$ 15 keV than for d + d fusion where the smallest Nτ occurs at ~100 keV.

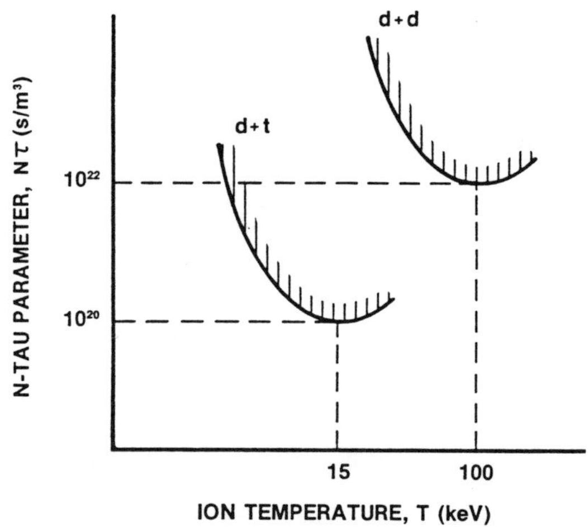

Fig. 10.4: Lower bound of the Lawson Nτ criterion for d + t and d + d fusion.

10.4 ICF Energy Balance

While the preceding analysis applies in particular to a pulsed magnetic confinement fusion (MCF) system, similar considerations can be employed for an inertial confinement fusion (ICF) system. The parametric characterization, however, has to be specific to the system of interest.

Consider Fig. 10.5 as a description for a laser or ion beam driven ICF system designed to compress a small pellet with the consequent release of fusion energy. Here, E_{in} and E_{out} possess a meaning similar to that used in the preceding section. Two special terms, together with the underlying processes are now introduced.

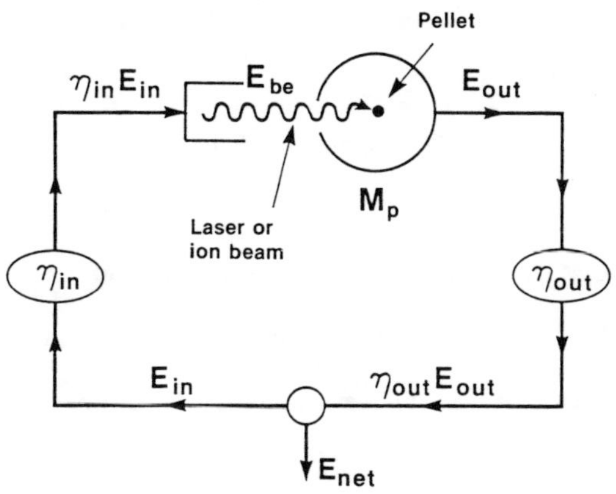

Fig. 10.5: Energy flow for an inertially confined fusion system.

The symbol E_{be} is to represent the laser beam or ion beam energy which leads to pellet compression and subsequent fusion burn. This beam energy is obtained from that supplied, E_{in}, according to an electrical-to-beam conversion efficiency η_{in} so that

$$E_{be} = \eta_{in} E_{in} , \tag{10.17}$$

Next we introduce the pellet energy multiplication M_p defined by

$$M_p = \frac{E_{out}}{E_{be}} . \tag{10.18}$$

Note that this energy multiplication contains the effect of fusion burn and consequent mass decrement leading to energy release; hence, M_p is expected to be greater than unity.

The overall station electrical energy, given by

$$E_{net} = \eta_{out}\, E_{out} - E_{in} \ , \tag{10.19}$$

can be expressed in terms of E_{be}, M_p, and η_{in} by a substitution of Eqs. (10.17) and (10.18).

$$\begin{aligned} E_{net} &= \eta_{out}\,(M_p\, E_{be}) - \frac{E_{be}}{\eta_{in}} \\ &= E_{be}\left(\eta_{out}\, M_p - \frac{1}{\eta_{in}}\right) . \end{aligned} \tag{10.20}$$

Requiring $E_{net} > 0$ demands that

$$\eta_{in}\, \eta_{out}\, M_p > 1 \ . \tag{10.21}$$

Knowing that $\eta_{in} < 0.1$, and supposing that $\eta_{out} \sim 0.3$ requires that the pellet energy multiplication exceed $M_p > 100$; indeed, the incorporation of other inefficiencies leads to the requirement of $M_p \sim 10^3$. The challenge therefore, is to design a pellet which responds to a given pulse so as to act as a most effective energy multiplier.

Problems

10.1 At what temperature will $P_{fu} = P_{br}$? You may use $A \simeq 10^{-38}$ for $P_{(\,)}\,[=]$ MW/m^3, kT [=] eV, and $N_{(\,)}\,[=]\ m^{-3}$.

10.2 Obtain an analytical expression for the extremum suggested in Fig. 10.4.

10.3 What are the principal differences for beam-to-target energy transfers between a laser drive and an ion driver?

Discussion / Analysis 10.2

Develop some general power/energy merit parameters for an arbitrary fusion system:

schematic:

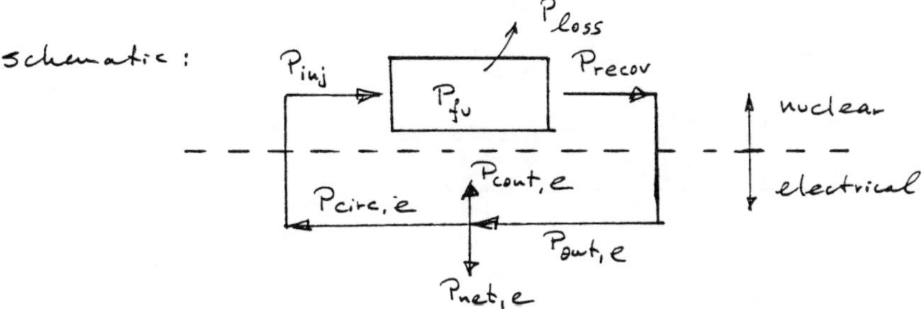

Efficiencies:

$$\eta_{all} = \frac{P_{net,e}}{P_{fu}} = \frac{P_{out,e} - P_{circ,e} - P_{cont,e}}{P_{fu}}$$

$$\eta_{elect} = \frac{P_{net,e}}{P_{out,e}} = 1 - \left(\frac{P_{circ,e} + P_{cont,e}}{P_{out,e}}\right)$$

Multiplications:

$$M_{all} = \frac{P_{out,e}}{P_{circ,e} + P_{cont,e}} \quad ; \quad M_{core} = \frac{P_{recov}}{P_{inj}}$$

To think about:

Both capital and energy investments are required before a plant can deliver power. Explore the "energy pay-back period" as a companion to "capital pay-back period" for existing power plants.

APPENDIX A

SELECTED DATA

Table A.1

Rest Masses

Electron mass:	m_e	=	9.1095×10^{-31} kg
Proton mass:	m_p	=	1.6725×10^{-27} kg
Neutron mass:	m_n	=	1.6748×10^{-27} kg
Deuterium mass:	m_d	=	3.343×10^{-27} kg
Tritium mass:	m_t	=	5.006×10^{-27} kg
1 u = 1.66056×10^{-27} kg			

Table A.2

Constants

Avogadro constant:	N_A	=	6.0221×10^{23}/mole
Boltzmann constant:	k	=	1.3806×10^{-23} J/K
		=	8.614×10^{-5} eV/K
Planck constant	h	=	6.63×10^{-34} J·s
Electron charge:	e^-	=	-1.60219×10^{-19} C
Permittivity constant:	ε_0	=	8.85×10^{-12} F/m
Proton–electron rest mass ratio:	m_p/m_e	=	1836.15
Speed of light in free space:	c	=	2.99792×10^8 m/s

Table A.3

Conversion Factors

1 eV = 1.602×10^{-19} J	1 tesla(T) = 1 weber/m^2 = 10^4 gauss (G)
1 u = 1.66056×10^{-27} kg	1 fm = 10^{-15} m
1 E/m = 931.50 MeV	

Table A.4

Neutron Cross Sections*: Selected Nuclides

Nuclide	Microscopic Cross Section (b, 10^{-24} cm^2)		
	σ_c	σ_f	σ_t
^{10}B	3840	0	3840
^{11}B	0.005	0	0.005
^{135}Xe	2.7×10^6	0	2.7×10^6
^{149}Sm	4.1×10^4	0	4.1×10^4
^{233}U	49	524	573
^{235}U	101	577	678
^{238}U	2.7	0	2.7
^{239}Pu	274	741	1015
^{241}Pu	425	950	1375

* v = 2200 m/s.

Table A.5

Selected Neutron Cross Sections for Elements

(Based on *Reactor Physics Constants*, ANL-5800 (1963), v = 2200 m/s)

Proton No.	Element	Material Density (g/cm^3)	Nuclei Density (10^{24}/cm^3)	Microscopic Cross Section (b, 10^{-24} cm^2)		
				σ_a	σ_s	σ_t
1	H	8.9†	5.3†	0.33	38	38
2	He	17.8†	2.6†	0.007	0.8	0.80
3	Li	0.534	0.046	71	1.4	72.4
4	Be	1.85	0.126	0.010	7.0	7.01
5	B	2.45	0.136	755	4	759
6	C	1.60	0.080	0.004	4.8	4.80
7	N	0.0013	5.3†	1.88	10	11.9
8	O	0.0014	5.3†	20†	4.2	4.2
13	Al	2.70	0.060	0.241	1.4	1.64
22	Ti	4.5	0.057	5.8	4	9.8
23	V	5.9	0.070	5	5	10.0
24	Cr	7.1	0.082	3.1	3	6.1
25	Mn	7.2	0.079	13.2	2.3	15.5
26	Fe	7.9	0.085	2.62	11	13.6
27	Co	8.9	0.091	38	7	45
28	Ni	8.9	0.091	4.6	17.5	22.1
29	Cu	8.1	0.085	3.85	7.2	11.0
36	Kr	0.0037	2.6†	31	7.2	38.2
38	Sr	2.54	0.017	1.21	10	11.2
40	Zr	6.4	0.042	0.185	8	8.2
41	Nb	8.4	0.054	1.16	5	6.16

† Value has been multiplied by 10^5.

Table A.5 (continued)

Proton No.	Element	Material Density (g/cm³)	Nuclei Density (10^{24}/cm³)	Microscopic Cross Section (b, 10^{-24} cm²)		
				σ_a	σ_s	σ_t
42	Mo	10.2	0.064	2.70	7	9.70
47	Ag	10.5	0.059	63	6	69
48	Cd	8.6	0.046	2450	7	2460
49	In	7.8	0.038	191	2.2	193
50	Sn	6.5	0.033	0.625	4	4.6
54	Xe	0.0059	2.7†	35	4.3	39.3
55	Cs	1.87	0.008	28	20	48
62	Sm	7.7	0.009	5600	5	5610
64	Gd	7.95	0.030	46,000	–	–
66	Dy	8.56	0.032	950	100	1050
72	Hf	13.3	0.045	105	8	113
74	W	19.3	0.063	19.2	5	24.2
79	Au	19.3	0.059	98.8	9.3	107
80	Hg	13.5	0.041	380	20	400
81	Ti	11.8	0.035	3.4	14	17.4
82	Pb	11.3	0.033	0.170	11	11.2
83	Bi	9.75	0.028	0.034	9	9
90	Th	11.3	0.029	7.56	12.6	20.2
91	Pa	15.4	0.040	200	–	–
92	U	18.9	0.048	7.68	8.3	16.0
93	Np	–	–	170	–	–
94	Pu	19.7	0.049	1026	9.6	1040

Value has been multiplied by 10^5.

Table A.6

Deuterium-Tritium Sigma-V Parameters

(Source: ORNL/TM-6914 (1979))

Temperature T (keV)	Sigma-V $<\sigma v>$ (m^3/s)
1	6.27×10^{-27}
3	1.81×10^{-24}
5	1.35×10^{-23}
7	4.14×10^{-23}
10	1.13×10^{-22}
30	6.65×10^{-22}
50	8.54×10^{-22}
70	8.76×10^{-22}
100	8.24×10^{-22}
300	4.90×10^{-22}
500	3.63×10^{-22}
700	3.02×10^{-22}
1000	2.55×10^{-22}

APPENDIX B

PARTICLE ACCOUNTING

The extensive use of mathematical methods as tools for the analysis of conceived physical phenomena is of profound importance in contemporary science and technology. One consequence of this union of two conceptually distinct intellectual inventions is the need for care in the interpretation of mathematical symbols as descriptions of physical processes. One area of broad interest in general and of particular interest herein is the so-called continuous-discrete dichotomy.

It is common to use the differential calculus -- which generally assumes "continuity" of certain functions -- to account for the time-variation of physical entities such as atoms, nuclides and nucleons -- which are "discrete" entities. For example, the dynamical equation

$$\frac{dN_i}{dt} = R_{+i} - R_{-i} , \tag{B.1}$$

describes the time rate change of the number of particles $N_i(t)$ in a unit volume which experience discrete reaction rate processes contributing to an increase (i.e. R_{+i}) or decrease (i.e. R_{-i}) in the population $N_i(t)$. Note that dimensionality considerations require that each expression for the reaction rates $R_{\pm i}$ must possess the same units as dN_i/dt.

At a microscopic level, the number of $N_i(t)$ particles varies by unit integer steps $(\Delta N)_{int}$ as a function of time described by time intervals of arbitrary length $(\Delta t)_{arb}$; a graphical depiction of this is suggested in Fig. B.1. Though mathematical techniques which describe the discrete processes implied in this figure do exist, they are nevertheless most cumbersome unless transformed in a manner which eliminates the discrete depictions.

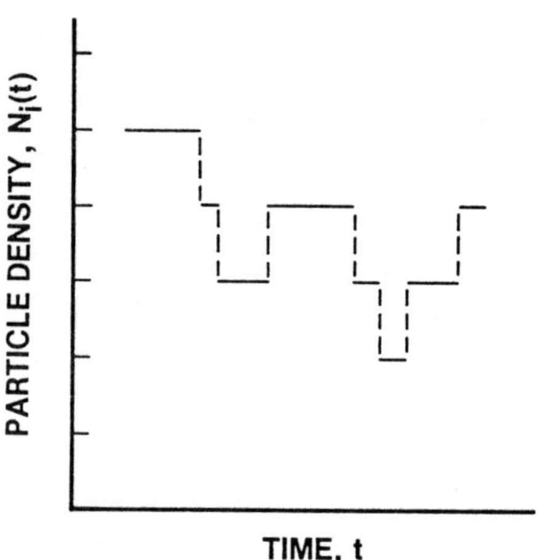

Fig. B.1: Depiction of the microscopic variation with time of the particle density $N_i(t)$; ($N_i(t) \rightarrow$ small).

Analysts have long known that methods need to be matched to objectives. If $N_i(t)$ were generally a small number, say 10 or 20, then an integer change represents a significant fraction and would need to be retained in detail. If, however, $N_i(t)$ were a large number, and here we are interested in quantities in the range of 10^6 to 10^{20}, then an integer change is of lesser importance and one is invariably satisfied with "smooth" trends of $N_i(t)$. An analogy with the population of a country is appropriate: while birth and death processes are taking place at perhaps fractional second intervals thus causing the last several population digits to change rapidly, few uses exist for such detail; indeed, national population data of general interest are normally rounded off to the nearest 10^5 or 10^6 for intervals of months and years.

In nuclear science, as in national or global population statistics, the magnitudes of the number of particles of interest are sufficiently large that, even for discrete (integer) transformations, a sufficiently "smooth" and hence differential variation with time of the population $N_i(t)$ is fully satisfactory. We suggest this in Fig. B.2. As a consequence, the left-hand-side of Eq. (B.1) is now defined by

$$\lim_{\substack{\Delta t \to 0 \\ N_i \to \text{large}}} \left\{ \frac{N_i(t + \Delta t) - N_i(t)}{\Delta t} \right\} = \frac{dN_i}{dt} . \tag{B.2}$$

While this justifies the left-hand-part of Eq. (B.1), some further elaboration of the right-hand-side is appropriate.

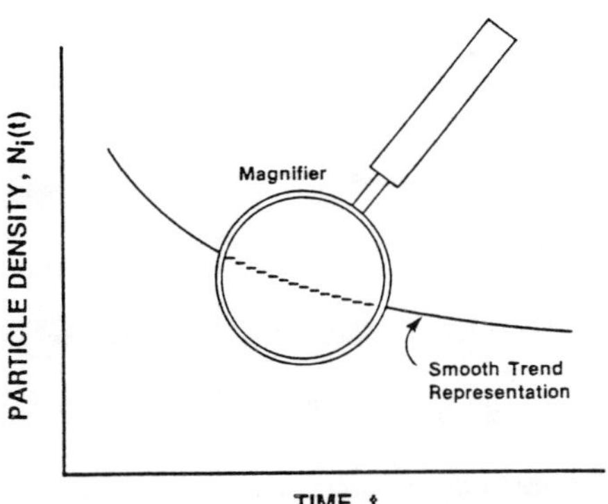

Fig. B.2: Depiction of the equivalent variation with time of the particle density $N_i(t)$; ($N_i(t) \rightarrow$ large).

The symbols R_{+i} and R_{-i} in Eq. (B.1) are labels for reaction rate processes -- that is "birth" and "death" rates -- which add to or subtract from $N_i(t)$ at any time t of interest. Appropriate algebraic expressions need to be specified for these symbols based on physically plausible processes of interest. The methodology employed in determining these expressions requires the use of relevant principles associated with various disciplines: e.g. mechanics, electromagnetics, probability, energetics, etc. Invariably, assumptions and approximations are invoked to arrive at useful expressions. Indeed, different approaches may lead to identical, similar, or even contradictory results; it is the analyst's experience and/or insight which are the important determining factors in this process of constructing mathematical analogues.

An example of such model building is provided by the dynamical equation which describes the number of nuclides $N_i(t)$when radioactive decay is the only reaction rate process. That is, we wish to determine R_{decay} in

$$\begin{aligned} \frac{dN_i(t)}{dt} &= R_{+i} - R_{-i} \\ &= 0 - R_{decay} \, . \end{aligned} \tag{B.3}$$

The principles commonly employed are the following:

1. The disintegration of any one nuclide from the population $N_i(t)$ in any short time interval is governed by a constant probability.
2. The fractional change in the population $N_i(t)$ in any short time interval is a constant.

Both approaches, which happen to be tautologies, yield

$$R_{decay} = \lambda_i \, N_i(t) \, , \tag{B.4}$$

so that the dynamical equation for Eq. (B.3) is simply

$$\frac{dN_i(t)}{dt} = -\lambda_i \, N_i(t) \, . \tag{B.5}$$

As a generalization, the most common reaction rate expression of interest, $R_{\pm i}$ in Eq. (B.1), can be reduced to the following three cases:

$$R_{\pm i} \begin{cases} = \text{constant} \\ = \text{proportional to some density } N_j \\ = \text{proportional to a product density } N_i N_j \end{cases} \tag{B.6}$$

A physically plausible justification for each of these processes can be proposed for conceivable conditions:

1. A constant production/destruction rate may be associated with external effects such as a device which generates or destroys particles.
2. A reaction rate which is proportional to some density is associated with, for example, radioactive decay or density dependent leakage.
3. A reaction rate which is proportional to the product of two densities is associated with a two-body collision (three-body collisions occur so rarely that they can be ignored).

Thus, Eq. (8.1) may therefore be written in a most general form as

$$\frac{dN_i(t)}{dt} = \pm R_i \pm \sum_j a_{ij} N_j(t) \pm \sum_k \sum_\ell b_{ik\ell} N_k(t) N_\ell(t) \tag{B.7}$$

Here, all algebraic symbols are taken as positive quantities and the plus or minus signs are chosen depending upon whether an increase (+) or decrease (−) in $N_i(t)$ occurs as a result of the reaction rate suggested by the term.

Note that all dynamical equations used throughout this text are but special cases of Eq. (B.7).

APPENDIX C

CROSS SECTION

The concept of a nuclear cross section has been introduced in order to specify various kinds of two-body interaction rates. Although theoretical tools exist for the determination of such parameters, the most reliable cross section information results from specially designed experiments. We describe here a particular type of experiment for such purposes.

Consider a collimated and mono energetic beam of nuclear particles, or any form of nuclear radiation, directed toward a sufficiently thin target, Fig. C.1. The target is here taken to be so thin that every atom in it is exposed to the incoming beam particles; that is, no atom shields another and one may associate an area σ_r such that every beam particle which, if it enters this area, will experience an r-type of interaction. This area σ_r is called the microscopic cross section for the r-type process involving the beam particle and the target atoms.

Supposing that the objective is only to measure the scattering cross section σ_s and that the target atoms are known not to engage in any other interaction with the incoming beam particles. Detectors strategically placed around the target will therefore measure only the scattered beam particles. By suitable summation, making corrections for scattering direction which do not contain detectors and other adjustments, it is possible to obtain the total scattering interaction rate R_s as

$$\begin{pmatrix} \text{Number of scattering} \\ \text{events in target} \end{pmatrix} = R_s \tau, \qquad \text{(C.1)}$$

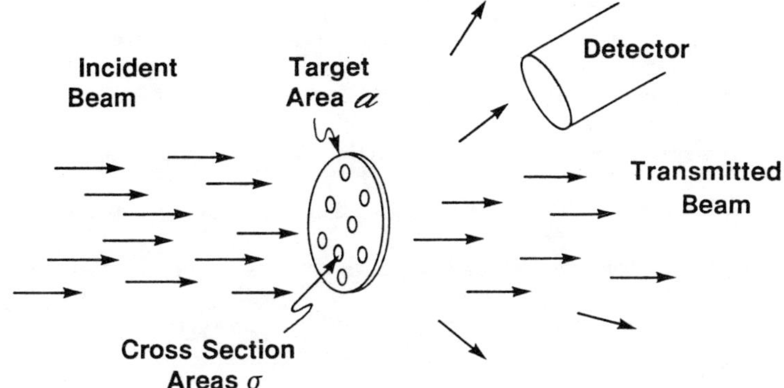

Fig. C.1: Schematic of experimental set-up for the determination of scattering cross section.

where τ is the measuring period. Similarly, with a known incoming beam ϕ_{in}, we write

$$\left(\begin{array}{l} \text{Number of beam particles} \\ \text{incident on target} \end{array} \right) = \phi_{in}\, a\, \tau , \tag{C.2}$$

where a is the beam area.

Knowing that only the fraction of the beam which enter all the individual areas σ_s will scatter, requires therefore

$$\phi_{in}\, a\, \tau \times \left(\frac{\sigma_s N_a}{a} \right) = R_s\, \tau . \tag{C.3}$$

Here N_a is the areal density of atoms in the target. We cancel common terms in this equation and solve for the scattering cross section of interest:

$$\sigma_s = \frac{R_s}{\phi_{in} N_a} . \tag{C.4}$$

With all factors on the right-hand side of this equation known by experimental design, the required cross section -- now specific for the process, for the target atom, and for the incident particle of a given energy -- has been experimentally measured.

It follows, of course, that different incident particles, different target atoms, different relative speeds between incident particle and target atoms, and for different processes, will all yield different cross sections. The collection and management of such data can hence be a very involved and complex activity. The cross section listed in Appendix A represents but a very small part of available interaction data.

APPENDIX D

EXPERIMENTS

Nuclear experiments invariably involve counting and recording of events associated with nuclear processes. The following represents a sampling of such experiments of relevance to the subject of this text:

1. Natural background radiation assessment.
2. Half-life of radioisotope determination.
3. Geometric and material shielding evaluation.
4. Nuclear cross section measurement.

A basic radiation detection apparatus is depicted in Fig. D.1. When a radiation photon or nuclear particle enters the sensitive volume of a detector, an ionization cascade in the gas is initiated to generate a current pulse. This pulse is then amplified and recorded. A timer specifies the interval τ during which counts are recorded, from which the count-rate R_d is determined; this count-rate is then related to the nuclear phenomena of interest. Note that R_d is actually a reaction rate occurring in the detector.

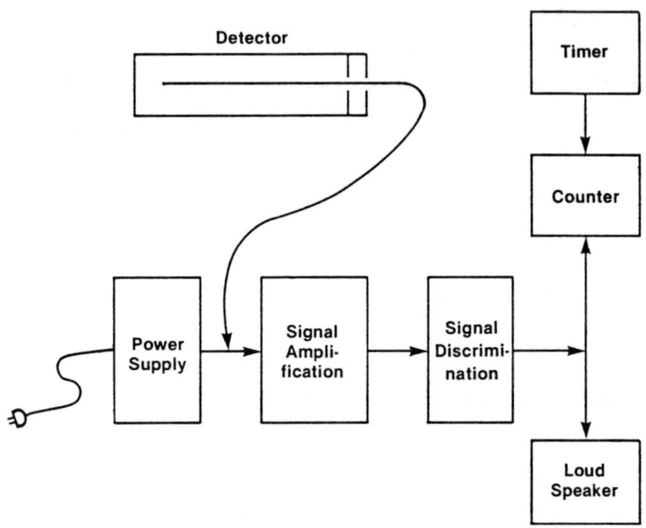

Fig. D.1: Block diagram of nuclear counting apparatus.

Discussion / Analysis D.1

Investigate the statistical variations of radiation emitted from radioisotopes.

Choose a suitable radioisotope (indeed, even the background radiation will do), specify a counting interval Δt, and collect a sufficiently large set of detector count rates $R_{d,i}$, $i = 1, 2, 3, \ldots$. The histogram might appear like the following:

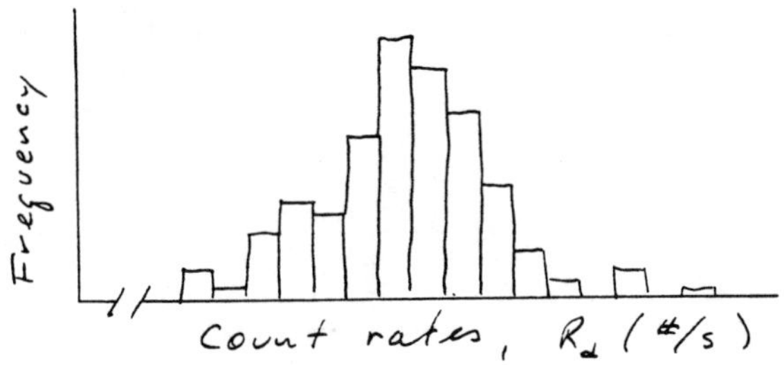

Compute : mean, variance
Perform : tests of symmetry, curve fitting.

To think about:
Might there be reasons to expect that some known distribution functions are applicable?

Discussion / Analysis D.2

Undertake a nuclear half-life measurement

Choose a radioisotope with a half-life comparable to the time available (see Chart of The Nuclides).

shielded enclosure

radioisotope contains $N_a(t)$

$$\frac{dN_a}{dt} = R_{+a} - R_{-a} = 0 - \lambda_a N_a(t)$$

detector provides $R_d \left(\frac{\text{counts}}{s} \right)$

Know $\quad R_d(t) = \eta (R_{-a}) \qquad \eta$ = overall efficiency

$$= \eta \lambda_a N_a(t)$$

$$= \underbrace{\eta \lambda_a N_a(0)}_{\text{some constant}} e^{-\lambda_a t}$$

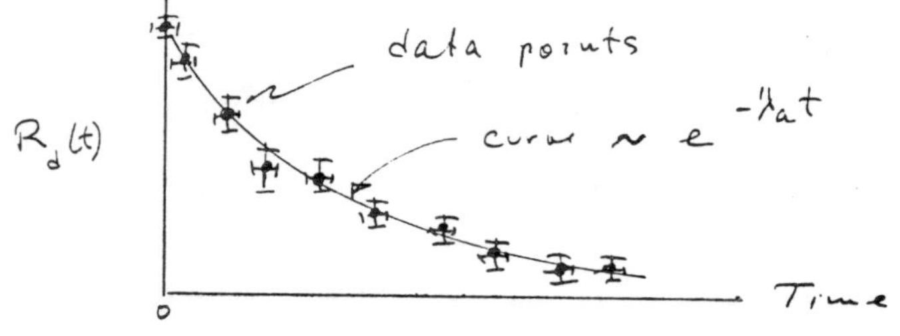

To think about:

Can you think of a transformation of the data to give $\tau_{1/2}$ as the slope on an experimental plot?

Discussion / Analysis D. 3

Undertake a radiation attenuation experiment.

Consider following:

$J_{in} = J(0)$ $\quad$ $J_{out} = J(x)$ $\quad$ Detector

0 $\quad$ x

For Δt specified

$$R_d \propto J(x) = N(x)\,v$$

R_d: detector count rate; $N(x)$: radiation density; v: speed of particles or photons

$$R_d(x) = \eta\, N(x)\, v$$

η : overall efficiency

$$= \eta\, v\, N(0)\, e^{-\mu x}$$

($\eta\, v\, N(0)$: some constant)

$R_d(x)$

data points

curve $\sim e^{-\mu x}$ (μ found from curve fitting)

0 $\quad$ Thickness

To think about:

Consider an experiment to test the geometrical shielding $1/r^2$ relation.

APPENDIX E

MATTER-ENERGY PERSPECTIVES

We had indicated in Chapt. I that matter and energy are the fundamental physical components of human existence and then, in the following chapters, we discussed and analyzed selected aspects of (nuclear) matter-to-energy transformations. In this Appendix, we point to a more broadly based integrative perspective of this subject.

The contemporary view of matter is that there exist fundamental building blocks which combine in numerous ways and at various hierarchical levels to form a range of material entities. We list one such classification in Table E.1 below.

Table E.1

Tabulation of Forms of Matter

Generic Name	Number of Forms
Photons (mass equiv.)	1
Leptons	3
Mesons	5
Baryons	8
Nuclides	$\sim 2\times10^{3}$
Atoms (incl. ions)	$\sim 10^{5}$
Molecules	$\sim 10^{8}$
Substances	$>10^{10}$

A companion listing for forms of energy does not appear feasible. However, we suggest a particular classification in Table E.2 which relates directly to the dominant matter-energy transformations of commercial and industrial interest.

Table E.2

Classification of Energy

Primary	Secondary
Environmental	Radiant (Sun)
	Air (Wind)
	Water (River, Tides, ...)
	Earth (Geothermal, ...)
Chemical	Wood
	Coal
	Oil
	Natural Gas
Nuclear	Radioactive nuclides
	Fissile nuclides
	Fusile nuclides

As a point of elaboration, note that while the nuclear category is clearly related to classes of nuclear structure rearrangements and similarly the chemical category characterizes rearrangements of molecular structures, the environmental grouping does lead to some overlapping implications. For example, the radiant energy from the sun -- which sustains such basic processes as biological life, photosynthesis and climatological processes -- is based on nuclear fusion reactions and hence constitutes nuclear radiation.

To this one may add many other problems of energy systematics which endow it with an enigmatic paradox of availability and of value: energy is both plentiful and scarce and also both cheap and priceless.

The historical evolvement of commercial and industrial matter-energy transformations on a global scale is characterized by some remarkable patterns of regularity. This is displayed in Fig. E.1 below which shows some matter-energy transformation time series on a scaled ordinate designed to highlight the similarity of market penetration rates. Thus, the recently dominant chemical energy technologies -- based on the hydrocarbon burning in coal, oil and natural gas -- required almost 50 years to proceed from 1% to 10% of the global energy market share. By these criteria, it would therefore seem that the recent growth rate of the nuclear (fission) energy technology appears too rapid.

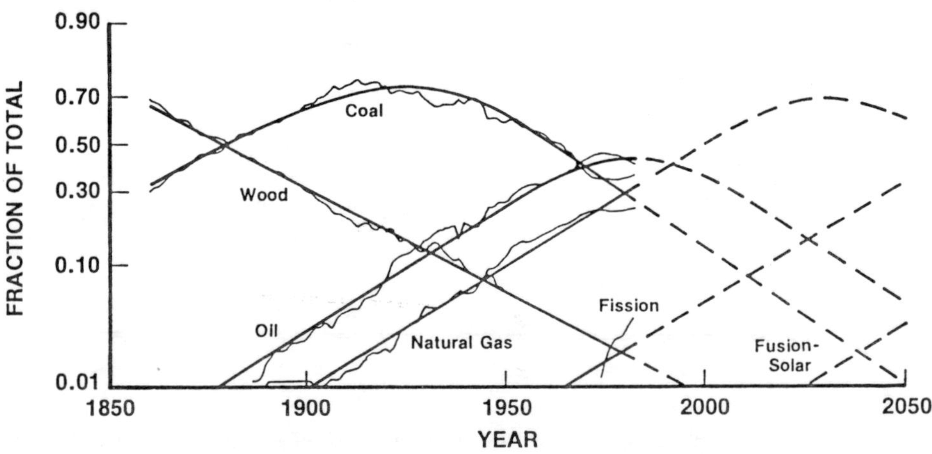

Fig. E.1: Time series of world primary energy substitution for 1850 to 1982 with extrapolation to 2030. (Adapted from W. Haefele, Energy in a Finite World, Ballinger Publ. Co., U.S.A.(1981) and updated by C. Marchetti (1986).)

This same figure also depicts a conceivable future global pattern of energy to be supplied by existing and emerging energy technologies; these extrapolations up to the year 2050 are based on expected regional growth, equitable availability of primary resources, and the intrinsic workings of a particular model of competition among various technologies (see reference in Fig. E.1 for further details).

Presently, some 400 fission reactors supply approximately 4% of the global energy requirements. As suggested in Fig. E.1, this is expected to increase during the next several decades. Then, towards the middle of the next century, a significant contribution is expected from fusion reactors and/or solar energy conversion devices.

On a time scale exceeding ~500 years and based on our current understanding of matter-energy transformations, it appears that the fusion of deuterium nuclides from the world's oceans represents the truly long-term and assured source of matter-for-energy.

APPENDIX F

SELECTED READINGS

1. J.A. Angela Jr. & D. Buden, Space Nuclear Power, Orbit Book Co., Malabar, FL, USA (1985).
2. G.S. Bauer & A. McDonald, Nuclear Technologies in a Sustainable World, Springer-Verlag, Berlin (1983).
3. J. Bernstein, Hans Bethe: Prophet of Energy, Basic Books, New York (1979).
4. R.W. Clark, The Greatest Power on Earth, Sidgwick & Jackson, London (1980).
5. R.W. Conn, "The Engineering of Magnetic Fusion Reactors", Scientific American, 249(2), p. 60 (1983).
6. F. Dyson, "Energy in the Universe", Scientific American, September, 225(3), p. 50 (1971).
7. H.P. Furth, "Reaching Ignition in the Tokamak", Physics Today, 38(3), p. 52 (1985).
8. A.A. Harms & W. Haefele, "Nuclear Synergism", American Scientist, 69(3), p. 310 (1981).
9. T.A. Heppenheimer, The Man-Made Sun, Little, Brown & Co., Boston (1984).
10. M. Taube, Evolution of Matter and Energy, Springer-Verlag, Berlin (1985).
11. A.M. Weinberg, et al., The Second Nuclear Era , Praeger, New York (1985).

INDEX

NOTES: